T0285348

READ THE TAPE

READ THE TAPE

USING SITUATIONAL AWARENESS
TO PREDICT BUSINESS
AND PERSONAL PROBABILITIES

MIKE S. SHAPIRO

Forbes | Books

Published by Forbes Books, Charleston, South Carolina.
Member of Advantage Media.

Forbes Books is a registered trademark, and the Forbes Books colophon is a trademark of Forbes Media, LLC.

Printed in the United States of America.

10 9 8 7 6 5 4 3 2 1

ISBN: 978-1-95086-361-7 (Hardcover)
ISBN: 978-1-95588-474-7 (eBook)

LCCN: 2023903207

Cover design by Joelle Hannah and Karen Weinberg.
Layout design by Matthew Morse.

This custom publication is intended to provide accurate information and the opinions of the author in regard to the subject matter covered. It is sold with the understanding that the publisher, Forbes Books, is not engaged in rendering legal, financial, or professional services of any kind. If legal advice or other expert assistance is required, the reader is advised to seek the services of a competent professional.

Since 1917, Forbes has remained steadfast in its mission to serve as the defining voice of entrepreneurial capitalism. Forbes Books, launched in 2016 through a partnership with Advantage Media, furthers that aim by helping business and thought leaders bring their stories, passion, and knowledge to the forefront in custom books. Opinions expressed by Forbes Books authors are their own. To be considered for publication, please visit **books.Forbes.com**.

In memory of my father, Lewis Shapiro,
who taught me that trying is succeeding, failure is a stepping stone,
humor tops everything, and unconditional love is invaluable.

And for Tara, my wife, sounding board, touchstone,
and partner in love and life.

CONTENTS

ACKNOWLEDGMENTS

Are any of us truly "self-made"? That certainly isn't the case for me.

As you'll see, this isn't a traditional "how-to" guide for professional or personal development. Instead, it includes lessons that people I've mentored tell me have been the most effective, presented in a novelized form and including experiences taken from my life. I want to acknowledge the people who shared these experiences with me.

My earliest (and most hilarious) failures and successes were episodes in which my family—dad, mom, and sister—played key roles. They supported my endeavors and taught me to keep failure in perspective, take disappointment in stride, and find humor in all. There aren't words to explain how empowering this has been.

I met my wife, Tara, when we were in college, and she's been the center of my universe since. She tells me the things I need to hear, and however things shake out, she sticks with me, often to my amazement and always with love. I'm fortunate and grateful.

I'd also like to acknowledge the mentors who I've had over the years—there are too many to name in a page or so, but you're out there, and your influence and insight stay with me every day.

And, finally, I'd like to thank all the people who have given me the privilege of mentoring them on their way to success—far and away, this has been one of the most exciting, satisfying, and humbling things that I could have hoped to do with my life. Thank you, all.

YOU ALREADY CAN

NEWPORT BEACH, CALIFORNIA

As Bennett Gates rolled up to the sleek recording studio for his noon session, he caught a glimpse of himself—or rather, a glimpse of his BMW iX reflected in the building's glossy black glass. For a split second, the image startled him. A year ago, he was driving a beat-up truck in the mountains of Virginia, feeling about as lost as he ever had. And now here he was, driving his brand-new BMW in sunny Newport Beach, wearing his favorite pair of bespoke leather shoes to his fancy noon appointment.

For most of his life, a question had haunted him. He had no doubt the Bennett Gates of a year ago would have had this question in mind when confronted with a shiny new BMW and the familiar face at the wheel: *Who does that guy think he is?*

Back then, that question would have loomed so large and made him feel like such an imposter that he might have turned his car around and started driving east. After driving all day and all night,

3

he would have pulled into the driveway of his childhood home, stuck his head in the sand, and hid from the world. Because at least when you were hiding, you couldn't fail.

Everything was different now. *I know exactly who that guy is,* Bennett thought, a smile spreading across his face, *because he's me.* He was learning to be comfortable in his own skin. Over the last twelve months, he had failed—spectacularly—but instead of being crushed by that failure, he had laughed and learned from it.

The irony, of course, was that only once he embraced failure had Bennett experienced such phenomenal success.

As he glided into a parking spot by the studio's front entrance, and Bennett took a moment to collect his thoughts, he couldn't believe he was really about to do this. A part of him was nervous. What if he said something stupid in front of Mike, his mentor? What if he couldn't find the right words to describe the transformation he'd experienced over the last year in his career, his confidence, his financial situation—heck, even his love life?

But that was just it. After everything he'd gone through, he was no longer afraid of failure. If he couldn't think of anything else to say to Mike, he would say that. And anyway, Mike would love it since that lesson on failure came straight from him.

As Bennett opened the car door, he saw motion at the front of the recording studio. He blinked.

If he wasn't mistaken, the five-time Super Bowl MVP had just walked out of the revolving glass doors. Bennett couldn't believe it. He rubbed his eyes to make sure it was really him. But there was no mistaking it. Everyone loved this guy, so Bennett had seen his face on his flat-screen TV dozens of times.

A black Mercedes-Benz G-Class with tinted windows pulled up to the curb. The man's assistant—or maybe his bodyguard?—jumped out to swiftly usher him inside.

Why am I surprised? Bennett thought. *I* am *at a recording studio in California, after all.* Still, he couldn't wait to tell Mike about his celebrity sighting. "I'm like a regular tourist," he said with a chuckle. He stepped out of his BMW, breathed in the salty seaside air, then pressed his key fob.

The studio's foyer was simple and elegant, with a few well-appointed pieces of furniture and modern art. One entire wall was made of glass, offering a dazzling view of the Pacific Ocean. As Bennett gazed out on the water, reflecting on how far he'd come, he felt a sense of peace wash over him. He was still getting used to feeling this way: at peace with himself and his accomplishments, while simultaneously being eager and even a little excited to share his experience with others.

"Is that Bennett Gates?"

He turned and saw his friend and mentor, who strode across the room and warmly shook his hand. It was only the second time Bennett had met Mike in person. The first time was under very different circumstances—they'd met on a soundstage to film a TV show, still one of the most mortifying moments of Bennett's life.

Mike seemed genuinely glad to see him, which made him feel good. He knew there were plenty of people the creator of the megahit podcast *Read the Tape* could be interviewing today, and yet Mike had chosen him, Bennett Gates.

"You won't believe who I saw in the parking lot," said Bennett.

"Wait. Don't tell me." Mike made a grand show of stroking his short beard. "The League's winningest quarterback?"

"Wait a minute," Bennett said, putting two and two together. "Are you telling me he was—"

"Here to see me?" Mike's face broke into a huge grin. "You bet. He launched a new business a few months ago, and he's a huge fan of the podcast. He said my sound bites have kept him going. The irony is that he gave *me* some of the best sound bites I've gotten yet. His episode's going to be great, wait and see. He set the bar very high."

"No pressure!" Bennett joked.

"Nah, no pressure. You're gonna be great." He beckoned. "Hungry? We ordered all kinds of stuff. You wanna eat?"

Bennett grinned. This wasn't the first time he'd been a beneficiary of Mike's hospitality, and his mouth was already watering.

"You bet," he said.

As Bennett surveyed the impressive spread, he thought again how feeding people truly did help them drop their guard. Eating together made conversations—and negotiations—flow more easily, something Bennett had personally witnessed just a few weeks ago. He couldn't wait to tell Mike all about it.

As Bennett ate, he *did* feel more at ease about being featured on the podcast. The tension melted out of his shoulders as he and Mike caught up on everything that had happened since they last saw each other. Amazing what a little food could do.

After Bennett finished eating, Mike gave him a tour of the studio, showing him the main recording area—a cutting-edge studio with thickly cushioned walls, platinum microphones, a full bank of monitors, and various state-of-the-art recording equipment. Along the way, Mike regaled him with an entertaining story of how he found this space with an ocean view and wasted no time in renovating it to suit his purposes. He had a big vision for the podcast, and he spared no expense in making the studio into something spectacular.

"You could call it *Extreme Makeover: Recording Studio Edition,*" he quipped.

"Mike always does everything grandiose," chimed in Tessie, Mike's cousin and the podcast's coproducer, who had slipped in the door and was walking over to greet them. "Let's just say the term 'humble beginnings' is not in his vocabulary. Good to see you again, Bennett. You look great."

"I *feel* great. I've been working out every day." Bennett chuckled. "Can you imagine me saying *that* a year ago? We just built a small gym annex at the office, and everyone on the team is loving it. Me most of all."

"Well it shows," Mike said. "You know, Tessie's the one who asked 'Why a podcast?' when I first had the idea. She always brings a little common sense to my wilder schemes. I said, 'You know why, Tessie. Who else do you know who likes to talk as much as I do?'"

Tessie laughed, her blonde hair bouncing off her shoulders. She gave Bennett a wry look. "He was right, of course. I don't know why I even asked."

"But I set the bar high," Mike said. "My mantra is 'finding joy in the success of others,' so the podcast couldn't just be me yammering on about how great and successful I am. I didn't want it to come across as some kind of cheesy, twelve-steps-to-success snake oil. We're so inundated with all this advice, all these people saying 'just do these four things' or 'take these seven steps and you'll be rich beyond your wildest dreams,' and it's just not applicable. Or if it is, well, yeah, you can do it, but does it make you happy and is it linear to who you are?"

Bennett nodded. Before he'd found his way to Mike's podcast— long before he'd set foot on the incredible journey that led him here, to this swanky studio in Newport Beach—he'd read plenty of books hawking some kind of snake oil for success. Or, more accurately, he'd read the first chapters, then never picked them up again.

He looked at Mike. It was hard to imagine his mentor had ever doubted that his podcast would be something special. Mike's one-of-a-kind brand of wisdom, humor, and advice had gotten him to where he was—running multiple companies, spearheading projects across the country, turning over billions in revenue every year. He'd built whole empires. What better person to help other aspiring entrepreneurs build their own?

Bennett knew his mentor well enough by now to know he hadn't gotten there easily. Mike wasn't one to shy away from challenges or obstacles, and he talked openly about his failures. And yet, somehow, he seemed to always arrive at the next thing with a smile on his face.

"You were the first person I felt really understood what I was going through," Bennett said. "I didn't feel like you were lecturing me or telling me what to do. It felt more like ... *you* were actually listening to *me*." He laughed out loud. "I know that sounds crazy, considering I was at home alone listening to a podcast of you talking."

> I talk about failure—because failure isn't an end point. *It's where you find your potentials and possibilities.*

Mike looked excited. "But that's the thing! That's exactly it. What I try to do is get people to listen to *themselves*, to focus on those things they enjoy doing and do well—how to interpret what's going on around them and use that to bring their ideas to life. The podcast was never meant to be a 'How to ...'; it's a 'You already can ...'"

"That's good," Tessie muttered to herself, reaching for her notepad. "That's *really* good, Mike. That could be the slogan for our next season." She looked up at them. "Guys, why aren't we recording? You're already dropping some real gems!"

Mike grinned at Tessie, then back at Bennett.

"By now you know I don't talk about success. I talk about failure—because failure isn't an end point. *It's where you find your potentials and possibilities.* So, Bennett,"—he gestured toward the empty chair—"are you ready to tell us about *your* failures? Think you can do that?"

A year ago, a question like that would have sent Bennett running for the hills; now, he just grinned. Tessie offered him a bottle of water and an encouraging smile. He accepted both as he settled into the chair.

"I don't *think* I can, Mike," Bennett said, sliding the headphones over his ears. "I *know* I can."

ONE YEAR AGO...

WHAT'S YOUR CHUTZMA?

SOUTHWESTERN VIRGINIA

Bennett Gates picked up a mud-smeared tennis ball and tossed it in a long, lazy arch across the field. Maurice, his silky terrier, leapt into the air and jetted after it, his tall, tan ears the only thing visible in the knee-high grass.

He was visiting family in this small mountain town because he wasn't sure what else to do with himself. Though after the last few hours, he was already regretting it.

It was six months to the day since he'd been fired from his job at a video game start-up. He still wanted to kick himself every time he thought about what had happened. Now that he had plenty of time to replay his career trajectory, he couldn't help but see the last few years as one failure after another, each domino falling because he'd made a mistake. Some mistakes were bigger than others, but they'd all led him to right here, right now—roving around this tiny town in the middle of nowhere, unemployed, uncertain of what to do next.

Bennett was a coder, and a decent one at that. A coder could always find work. But he had a bad feeling he wouldn't be able to work in video games again, not after "the incident."

The start-up was young and hungry—and they'd been stuck in beta forever, trying to debug their big multiplayer game before releasing it to the public. It had cycled through quality assurance so many times; Bennett joked that the game was "burning in the nine circles of QA." No one thought the joke was as funny as he did.

In the weeks leading up to Bennett getting fired, it was "crunch time," which was gamer code for "Get ready to work twenty-four seven." People were crashing at the tiny office in sleeping bags, and the two guys who founded the company—brothers with big dreams and not a lot of capital to fund them—were buying copious amounts of pizza and coffee to keep everyone fed and wired. The team was small, but the energy was electric. "This is it, guys," the older brother, who was CEO, kept saying. "We're about to change gaming forever. But we need all hands on deck. Every single one of us has gotta put in 110 percent if we're going to pull this off."

And Bennett just … couldn't. The game was okay, though he'd never loved multiplayer games—in his opinion, the story always suffered as a result. Plus he wasn't convinced it was quite the game changer his boss seemed to think it was. So he found himself dragging, stretching his lunch break longer and longer, sleeping past his alarm on the rare occasion he did go back to his apartment at night. Sometimes he let himself get swept up in the excitement. But even then, he'd talk himself down from the ledge. These guys were starry-eyed and naive, and they were about to get their asses handed to them. Bennett felt himself pulling back, not wanting to give 110 percent, or even 100. At the office, he'd start messing around with the game he was developing for his cousin, a silly pet project, when he was supposed to be working.

And then came the kicker: a slip of the tongue that would cost Bennett more than he ever could have imagined. On that fateful night, he'd left the office—which felt increasingly claustrophobic—to head down the street and grab a beer with one of the other coders. Three beers in, he was speaking freely—a little *too* freely, as it turned out. "Who thinks a multiplayer game is going to change the world?" he'd said, like a total idiot. "That's the dumbest thing I've ever heard."

The next morning, when he stood bleary eyed in front of big brother CEO and little brother CTO, it was Bennett's ass being handed to him, not the other way around.

"I'm not firing you because you said something stupid," said Bennett's boss, "though that certainly didn't help; I'm firing you because you're not bringing your whole self to the table, and you haven't been for a while."

It wasn't the first time Bennett had heard that tune. It was a new refrain to an old song. He packed up his desk—which amounted to his laptop and a neon-orange stress ball—and headed home. It was some consolation knowing that when the game did finally launch, it would get lost in the slew of other multiplayers. Bennett was gone before his stock options vested. But who cared? The company would fold within the year.

Wrong. The game broke every record in its very first week. Everyone was playing it: YouTubers, celebrities, presidential candidates. The start-up was all anyone could talk about, and the brothers launched to internet stardom instantaneously. Bennett lost count of the number of texts that lit up his phone, all saying something along the lines of "Hey, bro! You worked on that game, right? Congrats!"

He had made the biggest mistake of his life.

While everyone else's shares rocketed into the stratosphere—even the beer-slugging coder who had ratted him out was now a millionaire—Bennett lived off his severance check for the next few months, then moved in with his parents.

He was way too old to be living with his parents. He knew that. But until he found a job or something else to lift his spirits (and pad his pocketbook), he just wasn't sure what other options he had. There weren't a lot of opportunities for the guy who had snoozed through the biggest opportunity he'd ever been given.

Now here he was, six hours from his parents' home in the mountains of Virginia, in a hand-me-down beater truck from his dad. Bennett suspected his pops had only given him the truck because he felt sorry for him. At least it had given Bennett the chance to come visit his extended family, though—after the first four hours of interrogations by various family members on "Why'd you leave that video game company when you could have been famous!" and "When are you going to find a good girl?"—he was seriously rethinking that decision. After excusing himself from his middle aunt, who had badgered him with so many intrusive questions it might as well have been the Spanish Inquisition, he took off with Maurice and found somewhere he could be utterly, peacefully, alone.

That spot turned out to be a field, long gone to seed, on the very edge of town, where shops gave way to winding country roads and farmland. Together, he and Maurice took a long walk around the field, Maurice investigating every inch of the perimeter and Bennett taking in the fresh air and silence. Bennett dug an old tennis ball out of the truck's glove compartment and threw it out in the tall grass, where Maurice bounded happily after it. As he waited for Maurice to find his way back, he thought back to all those well-intentioned questions from aunts, uncles, and cousins.

What in the world happened to my life? he wondered. *How is it possible that I have continued to be my own worst enemy, failing time and time again?* Because the thing was, when Bennett cataloged his stumbling blocks, he always seemed to be the biggest cause of the stumbles.

He felt a soft thud against his leg. He looked down to see Maurice sitting patiently at his feet, his short tail, which curled up at the end like an elf shoe, wagging across the grass. Bennett threw the ball again, and Maurice leapt into action and retrieved it with simple canine grace. Then he trotted back, dropping his trophy proudly at his owner's feet.

"Good boy," Bennett said, scratching Maurice behind his pointy ears. *If only I could be that happy with so little,* he thought with a sigh.

At the sound of crunching gravel, he turned to see a brand-new Ford F-150 rumbling down the old farm road. A few minutes later, Jim Gates—his dad's youngest brother and the owner of Mill Creek Restaurant, a little country food dive just off of Main Street downtown—was ambling up to him. Bennett hadn't seen Jim yet on this trip, though he'd heard from his aunts that Jim and his wife Emma had moved since the last time Bennett was in town.

Jim had always been his favorite uncle. Bennett was closer to Jim than to his dad's two older brothers. When Bennett was a kid, Jim would take him on hikes and teach him everything he could remember about wilderness survival from his Boy Scouting days. Jim had two daughters—Marcy and Riley—so Bennett was like the son he'd never had, even if their "wilderness adventures" usually ended in several failed attempts to start campfires with sticks or to catch rabbits with bark fiber lasso traps.

But that never mattered to Bennett. What he liked most about his Uncle Jim was that he would let him talk—about life, dreams, girls, whatever was on his mind. And he'd offer advice if Bennett asked for it or tell his own story that usually blew Bennett's out of the water. But Bennett always felt comfortable around him, so even though he desperately needed a break from the nonstop visiting and small talk and questions about his personal life, he was happy to see him.

"Son, you are about as easy to track down as a polar bear in a snowstorm," said Jim, his expansive smile still visible behind an impressive handlebar mustache.

"Can't blame a guy for trying," Bennett said as he shook his uncle's outstretched hand. "How did you find me?"

"Alan Rector up at the seed store saw your truck turn down this road an hour or so ago. I figured you might be trying to get away from family for a bit. Those aunts of yours ask more questions than a House subcommittee." He winked at Bennett. "Didn't work as well as you hoped, though."

Bennett rubbed his hair and offered a halfhearted smile. "No, it's okay. I honestly didn't know where else to go. I probably should have come down to see you at your restaurant, but I just needed this"—he gestured to the wide-open sky and rolling green hills in the distance—"for a minute." He nudged the tennis ball away from Maurice's determined paws and tossed it again. "Everyone just keeps asking me what I'm going to do with my life."

Jim nodded and let Bennett's statement hang in the air for a moment.

"Do you?"

"Do I what?" Bennett asked.

"Do you know what you're going to do?"

Bennett slowly shook his head. "No, I don't. It's like I take one step forward, two steps back, and the whole time, wherever it is I'm trying to go just gets more ambiguous and further away. I've made so many mistakes, Uncle Jim. Life-altering, career-ending mistakes."

"Like with the game, you mean?" Jim asked.

"Yeah, with the game. But that's just the latest in a long line of failures. I have no home, no aspirations, no real love life to speak of … I feel like there's something out there that I could do, that I could be successful at. That I could actually enjoy doing! But I sure didn't find it at my last job, or even the one before that. So if it's out there …"

He trailed off for a moment and watched the grass sway as Maurice got sidetracked from the tennis ball, likely in pursuit of some small, unsuspecting creature.

"What happened at the last job?" Uncle Jim asked. "The one before the game company, I mean."

Bennett studied him. "You really want to know?"

"I wouldn't be asking if I didn't."

Bennett let out a long, belabored sigh. "You remember my college buds. Great guys, right? When they wanted to start a company together after we graduated, I said yes. I was the only one who could code, so I jumped on board. And it was fun for a while, the four of us just hangin' out, dreaming about what the business could be."

Jim nodded. "I remember talking to your dad about it. He was worried you were just killing time, stalling out on your potential."

"He wasn't wrong. The problem was, my college buds—they didn't have a lot of business savvy. It took a while for me to realize that. They were great guys, just not great business partners. But I was comfortable where I was. They never asked me to try new things, never pushed me too hard. So I stayed in my comfort zone. I stayed in that zone for *years*. Years longer than I should have, truth be told. I knew the company was headed nowhere fast, but I was too much of a coward to let go and move on."

"But then you got recruited," said Uncle Jim. "Right?"

"Did you already hear this story from my dad?"

"Only the rough outline, not the juicy bits. Go on."

Bennett sighed. "I got headhunted—by a tech company you've definitely heard of, a company anyone in their right mind would have said yes to. And do you know what I told the recruiter?"

"I'm not gonna like the answer, am I?"

"I said, 'No thanks. I'm good where I am.'" Bennett's laugh was rueful. "I chose to stick it out with my college buddies, because I was coasting. I think deep down I was scared of going to a big-deal company and just … not being good enough. Better to be a big fish in a small pond."

Bennett swallowed hard. "Joke's on me. A month after I talked to that recruiter, the pond drained. My friends called it quits. Shut down the company and went off in their separate directions, two to get an MBA and one to law school."

Uncle Jim frowned. "Why didn't you call the headhunter back and say you'd made a mistake?"

"Oh, I did. I called them up immediately and said things had changed and I could take the job. But they told me they'd filled the position. Actually, what they said was, 'You hesitated, so we moved on. If you're all in, you're all in.'"

A shadow passed over Bennett's face. There it was, that same refrain. *If you're all in, you're all in. You're not bringing your whole self to the table, and you haven't been for a while.*

"Damn, son," said Uncle Jim. "So you stayed too long with the guys who weren't going anywhere … and bailed early on the guys who were."

"Yeah," Bennett said, hating that it was true. "Something like that."

"Can I ask you a question, Ben?"

"Shoot."

"Do you still enjoy doing the work? I know you've wanted to work in computers since before you could pronounce the word 'developer.' But I can't say I've seen you really love it, or get really excited about making apps, for a long time."

Bennett considered it. "I still enjoy the work. At least, I think I do. I'm not so sure anymore. These days, apps feel like a dime a dozen. There's something like a thousand new ones published every day. And yet I still feel like I could create something, an app that really matters. Something that helps people. I used to get so fired up at the thought of developing an app that could change people's lives. But I haven't done it yet. I've just been burning time, making other people's apps or video games, and nothing ever takes off. Or the one that did take off is the one I didn't even like that much. And now I can't even get a second interview with any of the start-ups I've talked to."

He raised a quizzical eyebrow at his uncle. "Be honest, is it my jawline? I'm a little asymmetrical, I'll admit …"

Jim laughed so hard that Maurice stopped midpursuit and raced back to Bennett's defense. After a quick sniff check and a serious side-eye at Jim, he leapt back into the grass to find his prey.

"You've got every inch of my big brother's gangly height and mop-brown hair," Jim said. "I'll give you that. Would it kill ya to put a little muscle on those bones? But, son, folks don't hire based on whether you're magazine quality or not, especially if you're glued to a computer all day."

Bennett frowned. "I'm just at a loss, Uncle Jim. Maybe I should give up on finding something I'd enjoy doing and go code for some big company, where it's less about innovation and more about typing in ones and zeroes. Be content with a steady paycheck and leave it at that."

Jim tugged at the corners of his mustache for a moment, then put a hand on Bennett's shoulder.

"I'm not going to tell you what to do, Ben. Get a crappy desk job. Fine. Put in your eight hours, and find a hobby to make up for the soul you lost doing all that mindless paperwork. But do you really want that to be your life? Don't you want to be happy?"

"Of course," Bennett started. "But—"

"No *but*," his uncle interrupted. "You either want to be happy or you don't."

"Then yes. I do want to be happy. I just don't see how."

"That's a start."

His uncle pulled his phone out of his jeans pocket and scrolled for a minute, and a moment later Bennett heard his own phone ping.

"I want you to listen to this. A good friend shared it with me a while back, and it got me thinking about things differently, in a good way. It's the kind of thing I wish I'd heard when I was your age, but I'm glad I heard it when I did." Uncle Jim put the phone back in his pocket and patted Bennett on the shoulder again.

"I've got to run. Do yourself a favor and listen. It won't make you better looking, but it might change how you look at this whole job situation."

Bennett nodded. "Okay, yeah. I promise."

"Good." Jim bent down and ruffled Maurice's long fur, which Maurice graciously allowed. As he did, Bennett couldn't help but notice part of a watch peek out from under his uncle's sleeve. *Was that a Rolex?* he thought, astonished. Couldn't be. Last he remembered, his uncle's business wasn't exactly a moneymaker—just standard country fare with barely enough of a profit margin to get by.

"I'll check in with you later, handsome," Jim said with a laugh, and with that he was gone, rumbling down the road and back toward civilization.

Bennett threw the ball a few more times, until the time he threw it far enough that it disappeared into some brush. Maurice was flummoxed, running in circles, unable to fetch his favorite toy. When Bennett jogged over to the brush to see if he'd have better luck, he couldn't believe how winded he became. Really? Two minutes of jogging, and his lungs screamed in agony? He really needed to start exercising more. And by "more" he meant he should really start exercising, period.

Sadly, he couldn't find the tennis ball. It had been swallowed up by some swampy terrain at the far end of the field. Bennett sighed.

"We'll get you another ball, boy," he said, though Maurice looked disappointed. Who *wasn't* disappointed in Bennett these days?

He picked up Maurice and made his way back across the field— more slowly this time—and hopped into the truck. Then he revved up the engine and headed down the long, winding road, back to where he was staying with his grandmother.

It wasn't until hours later, after a hearty meal and a long porch chat with the sweetest little gray-haired lady south of the Mason-Dixon Line, that he finally opened Jim's message.

"Read the Tape," it said.

It was a link to a podcast. Bennett scanned the series of episodes, with odd titles such as "What's Your Chutzma?" and "All This Work Had Better Be Fun."

Why not? Bennett thought. He got comfortable on the old foldout couch where he always slept when he visited, tugged what little bit of blanket he could out from under a surprisingly heavy and happily snoring Maurice, and began to listen.

Welcome to Read the Tape. I'm your host Mike Shapiro, an author, entrepreneur, and venture capitalist. I've developed and created multiple businesses and done lots of things in my life. Honestly, a lot of my success has come from the failures I've had before. I am a big proponent of failing and learning from it ... pivoting, getting up—and moving forward. On this podcast, I'll be sharing stories of my own failures. Trust me when I say that I've had a lot.

Today we're going to talk about chutzma. Ever heard of chutzma? It's one of my favorite Mikeisms. Mikeisms are the things I've learned from a lifetime of experience that I try to quantify in my life and business so that you can put them to use in yours. After today's episode, you'll know exactly what chutzma is—and why you need it.

I want to start by telling you the story of when I played cello in the school band. The funny thing is, it's actually the story of when I didn't play cello at all.

Bennett sat up a little straighter on the old foldout couch. He had to admit he was intrigued by Mike Shapiro. There was warmth and humor in his voice; Mike sounded like a guy who didn't take himself too seriously. But there was also an underlying confidence and authority to everything he said. He sounded like someone Bennett wanted to know.

I was such a bad musician. That's not me being modest. As a kid, I took piano lessons. I was awful.

But my elementary school had a requirement that we all had to play an instrument. So I picked the cello, an instrument that was bigger than I was.

My sister chose the harp. So there we were, two little kids schlepping a giant cello and a gigantic harp around. Our parents were really laughing. Chalk it up to yet one more amusing story of our growing up.

Like so many other things I've tried over the years, it was immediately apparent that I was not good at it. So there I was in the school orchestra, without a musical bone in my body. Yet despite the fact that I was a total train wreck, they never kicked me out. Why? Because once I realized that a future in music wasn't in the cards for me, I embraced the experience of trying.

Not trying to be a musician.

Trying to be an *actor*.

I was so bad that the band director told me, "Please do not ever let the bow touch the strings."

Some other kids might have been devastated to hear that from their teacher. Not me. Growing up, I don't remember getting any positive accolades from my parents for anything I did successfully. Whatever the situation, whatever I did wrong, whether I failed miserably at being athletic or academic or anything else, I only remember my father on the sidelines yelling, "That's my kid! That's hilarious! That's fantastic!"

Our parents taught my sister and me to look for possibilities and potential—not end points—in whatever we did. They also encouraged us to find humor in our missteps and embrace failure. Their lessons helped me achieve success beyond anything that I could have imagined years ago, and for that, I'm incredibly grateful.

So picture me, this little kid, and this giant cello—a cello I had *no* idea how to play. For almost the entire year of indentured musicianship, I "played" that cello, furrowing my brow in concentration, bringing my bow back and forth, back and forth—and not once did I let the bow touch the strings.

Other kids' parents might've been disappointed. But mine embraced the ingenuity. My father thought it was hilarious. He announced to the entire audience, "My kid can't even touch the bow to the strings!" And get this: I didn't feel embarrassed. I learned I was bad, and that was okay. I was still down in front. In fact, I probably got *more* attention, which was fine by me.

We were all grateful that the instructor let me fake my way through something that clearly wasn't in my skill set. After all, what was the harm? It was an extracurricular activity. I wasn't being graded, and I wasn't bumping anyone from having the same opportunity.

In fact, there were two other empty cello seats the entire time I was there.

I'd found a way to use my failure at one thing (playing music) to hone success in another (acting), and my instructor was simply relieved that—if I wasn't going to throw in the towel—at least I wasn't going to bring the whole orchestra down with me.

From that perspective, everyone won. Yeah, it was a little brazen, but it was the first time I remember turning a negative experience into a positive one. And I enjoyed it.

That's the earliest memory I have of exerting chutzma, which is what this episode of "Reading the Tape" is all about. Chutzma. What it is, why it's important, what you can do with it, and how I may have made up the name but the idea is as old as dirt. It's the best way I know to find joy in what you do.

Bennett was intrigued. He'd never heard a story told quite like this. A story about someone failing—and taking pride in it. Mike wasn't sharing data he collected or abstract theories he'd concocted based on research and interviews. No, he sounded like he was living what he talked about, and that energy came through in a way that piqued Bennett's curiosity. Who *was* this guy?

Now, what chutzma looks like for me will be different than how it manifests with you. That's a given, because we're not the same person; our experiences, interests, goals, fears, obligations, and dreams are all different, even if they look similar on paper.

But *chutzma* is essentially a combination of two things: chutzpah and charisma. For those who aren't familiar with Yiddish, *chutzpah* is "self-confidence." When you bring those elements together, chutzma becomes the practice of developing a deep belief in yourself and your capabilities, strengthening that belief, making yourself stronger *through* that belief, and getting others excited and engaged in you and your cause.

> Developing chutzma doesn't require you to change who you are but rather to figure out how to bring out the best parts of yourself.

Developing chutzma doesn't require you to change who you are but rather to figure out how to bring out the best parts of yourself and use those to your advantage.

So let's start with charisma.

Everyone has a charismatic quality, whether they know it or not. I call this your *inner genius*. That's another one of my Mikeisms, and I really believe in the concept. Every single one of you has an inner genius, something that makes you special, sort of like your fingerprint. Something about you is very special, and you need to find out what it is.

You know the bullshit that's in all these books about how to become successful? I coach completely differently. I look at who you are and what makes you special—and I enhance *it*. *That's* your inner genius.

Another way to ask this question is, What's easy? In your entire existence, what was just really easy for you? Think back to elementary school. What part of your life was simple? Was it easy to make friends, be in math class, write? What is it about you that was super easy?

That's where I begin to search for this inner genius so we can find what works for you, because *that* is what makes you the most successful.

Bennett frowned. It troubled him that he didn't have an answer right away. When was the last time *anything* had felt easy? And yet, at the same time, he felt a stirring deep inside him, a little voice he'd been neglecting. He *did* have an inner genius. Somewhere. He just had to remember it.

He zeroed back in on the sound of Mike's voice.

Even some animals have figured out how to use their natural charm to make their way through the world. In fact, I can think of one right now: Maxine.

I've always been a dog person, and my wife and I have had several over the years, but Maxine is one of a kind. A Yorkie who weighs all of five pounds and fits in the palm of my hand, Maxine has an amazing ability to bend people to her will. Whenever she is threatened or needs attention—or, more often than not, when she just wants a treat—she will perform what has come to be known as the "Maxine maneuver." You can almost anticipate it happening. She suddenly stops whatever she is doing, flops onto the floor, rolls on her back, and sticks out her tongue.

Everything changes when she does that. Other dogs back off. If she is being reprimanded, we stop. Or if we are ignoring her, she instantly has our attention. Any time she needs it, she drops into her pose, and almost every single time, she gets what she wants, instinctively using her small size to her advantage. Maxine has found a way to channel her inner genius. Because of that, she has more charisma and confidence than many other big dogs I've met—and even some people.

If a scrappy little dog can figure out how to use her apparent disadvantages to her benefit, shouldn't we all be able to figure that out? Yes, we can. It's simply preparing ourselves for success through our attitudes, beliefs, and actions and knowing what we want so we can figure out how to get there.

Charisma also naturally comes through when we're doing what we love, and that could be anything. It could be that you love making other people laugh, or you find deep satisfaction in building a complicated scale model. You may find happiness in the human interactions of online gaming or in the solitude of writing a poem or painting. The point is that there has to be something, *something* you do that you not only enjoy but are naturally good at. Something that makes you special. Something that gave you joy.

So what is it?

In the questioning silence, Bennett tried to remember the last time he felt true joy. It felt so distant and unattainable, like a long-ago vacation he'd long since forgotten. Then Mike's voice came back, almost like he could read Bennett's mind.

And don't tell me that you've never found joy in anything.

Everyone has an inner genius. There is something in your life that you've done that you felt good about, one positive thing that's happened in your life. Something that felt easy and fun. Think about it. I'll wait.

The sound of dead air filled the room as Bennett stared at the old popcorn ceiling, pondering.

What brings me joy?

At his feet, Maurice snorted and twitched a paw as though chasing after a dream squirrel. And then, suddenly, Bennett felt a jolt of memory. Maybe it was Mike's story about Maxine. Or maybe it was that quiet voice deep inside himself that he hadn't listened to in a while.

Earlier that year, Bennett's twelve-year-old cousin, Todd, had gotten sick. He'd eventually made a full recovery, but for a few weeks there, it was rough. Todd had been out of middle school, mostly bedridden, for almost a month. Normally Todd's favorite thing to do was play with his golden retriever, Daisy, and take her on long walks around the neighborhood. So the fact that he had to stay in bed to rest for a month, and couldn't walk Daisy at all, had been hard.

On a Zoom with his cousin, Todd talked all about how much he loved his dog—which was of course something he, Bennett, could relate to—and about how much it was killing him not to be able to walk Daisy. Right then and there, an idea sparked in Bennett's mind. He decided to develop a custom app for his cousin that used actual video footage of Daisy and the streets around their house, creating a sort of virtual walking-the-neighborhood experience. The app would allow Todd to not only walk Daisy but throw her a virtual ball, watch her chase virtual squirrels, and even give her avatar a belly rub.

The timing wasn't ideal. It was right during crunch time with the multiplayer game, and Bennett didn't need any distractions. But he couldn't help it. He was pumped about Todd's game. He drew on the game design courses he'd taken in college, back when he was a huge gamer himself. Working on Todd's app gave him a way to hone his skills in basic UI design principles: visual and functional consistency, typography, and hierarchy (i.e., determining which elements on the mobile screen should carry the most visual weight). While coding had always come naturally to Bennett—that part of his new project was easy—it surprised him how much he enjoyed tinkering with the elements of design. Often he would start working on the app at 6:00 p.m., then look up and realize it was midnight. While he was engaged in something he felt passionate about, the hours flew by.

But even more than that, Bennett enjoyed creating something that so clearly filled a need. The multiplayer game he was supposed to be working on at his job just didn't move him. The game seemed no different from all the other ones on the market. Bennett knew it was silly—who made an app where the game play was walking your avatar dog and teaching it tricks?—but when he worked on Todd's game, he felt like the work he was doing really mattered. His cousin was heartbroken that he couldn't be out in the world with Daisy, especially after the doctors told him it might take months for him to fully recover. And suddenly, with the app, Todd would get to have Daisy be a part of his day whenever he wanted, even if he couldn't hop out of bed and take a stroll around the neighborhood.

Bennett loved making people happy. He'd listened to what Todd said he needed and filled that need. He had actually been so hyped about making the app he'd pulled a couple of all-nighters—no one at work batted an eye, of course, since it was crunch time and they assumed Bennett was working on their game—and he presented the app to his cousin a week after their conversation. It looked a hell of a lot better than the janky games with 1980s-Atari-style graphics he'd designed in college, and it was frankly the best-looking app he'd ever designed. It still had a couple of glitches, sure, but he could fix those in the updates. That had always worked for Steve Jobs.

Todd was thrilled. His parents said that he spent hours on the app and they'd hear him in his room, laughing away, having a grand old time. Real-life Daisy would curl up beside him in bed, while virtual-avatar Daisy roamed the neighborhood, chasing squirrels. Both Daisys got ample belly rubs—and Todd recovered far more quickly than the doctors expected. Aunt Debbie and Uncle Bo had always thought Bennett was the reason he bounced back so fast.

"You gave Todd exactly what he needed," Debbie had told him tearfully. "I believe your app is what helped him get well."

For Bennett, the whole thing had felt easy. Effortless. Fun. It was a simple app, silly even, but it had been a perfect use of his time and talents, and it had brought him so much joy.

Joy. I felt joy.
Suddenly, Mike's voice was back on.

> Did you think of something? Great. And if not, that's okay—it's your "homework assignment."
>
> So now that your wheels are turning, let's talk about the second part of chutzma: self-confidence.
>
> Going back to the story of my cello performances ... So I never actually played the cello. But my parents were proud of me, I was getting my music requirement out of the way, and I was doing something I enjoyed: acting.
>
> That willingness to stand out, to be noticed, is not easy. And even when people do choose to step into the spotlight, they may not necessarily be acting true to themselves in that moment. Because it's easier, sometimes, to be someone you're not and to assume that any criticism that comes your way isn't directed at *you*, just the false perception of you.
>
> But that's not chutzpah. Chutzpah is about *self*-confidence. It's about being comfortable in your own skin and confronting those irrational fears that keep you from taking a chance.
>
> Say you're at a café, and you see someone you admire or a senior at your company you've been meaning to introduce yourself to. You want to walk up and say something to them, but suddenly those irrational fears manifest:

What if they don't like me?

What if I stumble all over my words? What if I have nothing to say?

What if, what if, what if …

Well, what if you just walked up and said "hi" and introduced yourself? What's the worst that could happen? You'd be embarrassed? Maybe. But is that really all that bad? The fact is—and you're going to hear me say this a few times over the course of these episodes—you only need to be right 51 percent of the time to be successful. If

> You only need to be right 51 percent of the time to be successful.

you're wrong 49 percent of the time, so what? It's better to be comfortable with that failure because, in the end, you're still coming out on top. In the meantime, life is so much easier when you learn to laugh and roll with the punches.

A while back, I was mentoring a fitness influencer in his thirties. We usually met up at the gym, which worked great for me—the gym is basically my second home. Fitness is incredibly important to my daily experience, and I encourage you to find a way to make it important to yours. I have a trainer, but you don't have to have a trainer to exercise. Try jogging, walking, whatever it takes to keep moving. We'll talk more about that later.

So one day, we're in the gym, and this guy's idol walked into the room. He was huge, a major body-builder who was well known in professional circuits, and my mentee just froze.

"Why don't you go say hi?" I asked.

He shook his head. "I'm scared to talk to him."

Now, this guy was no wimp. He was almost the same size as his hero, but just the idea of walking up to him had my guy caving into all his "what ifs."

So, like a good mentor, I helped.

"Hey!" I said, walking up to this guy who has certainly thrown a few weights around. "So you're the most famous guy here in bodybuilding."

"Yeah," he said.

"Well, this other guy wants to meet you."

He was embarrassed and caught off guard, and yet, a minute later, they'd struck up a conversation and were sharing contact information.

It's hard to stop being fearful of putting yourself in the spotlight, but the only way we're going to learn who we are and how to be comfortable with ourselves is by facing those irrational fears head-on.

In other words, to borrow a very well-known phrase, "Just do it."

Easy for you to say, Bennett thought. He could recall half a dozen times just in the past few days when he chose to simply avoid eye contact rather than walk up and speak to someone. With his nosy aunts, he actually wished he had.

But again, Mike had a point. What's the worst that could happen? Bennett hated feeling embarrassed. But was he giving more credit to that feeling than it really deserved? What if he just walked up and said "hi"? What if he let people see him for who he was rather than who he tried to be?

That question brought Bennett back to his earlier question: *What brings me joy?*

Again he thought of Todd's face when Bennett had installed the app on his cousin's phone. He thought of how grateful his Aunt Debbie was—how she truly believed Bennett's silly app, which had only cost him a few days and dollars to make, had helped Todd heal. It was one of the only times in Bennett's life he felt like he'd been really good at something, *and it had helped someone.* It was a feeling like no other. At the time, it had felt like, well, a kind of genius. An inner genius, you might say.

He hadn't felt that in a long time. Certainly not at his old job where he'd felt mostly uninspired.

Now Bennett realized he missed that feeling. He missed it a lot.

But Mike wasn't done yet. Bennett was starting to get the sense that this guy always had more to say.

Here's another thing about chutzpah and charisma: they're driven by communication. That communication may not be verbal, but both are an exchange, on some level, of your true self. And when you're yourself around others and not trying to be someone you're not, that's evident. People are naturally attracted to those who are genuine and aligned with their true self—who are transparent and honest and capable of accepting both their faults and what they've done well.

So what's your inner genius? What brings you joy? And how can you use that joy to build up your self-confidence, your chutzpah?

Are you eccentric? Your chutzma could be enchanting people with your quirks.

Meek? Lead by letting others take the reins.

Introverted? Shine with understated elegance.

Bookish? Turn smart into an art.

No matter how you perceive yourself to be, that *chutzma* is hidden within. When you're ready to tap into it, you will find the success you're looking for.

Now, I mentioned homework earlier, and I'm going to add to that. I asked you to take some time to think about those things that bring you joy. Write down one of those things. Make it a good one. And while you're doing that—or maybe to help you as you do that—I also want you to write down five things that you're *not*. We'll talk about those more in the next episode. Because here's what I'm really getting at with all of these questions:

Who are you?

More on that next time.

Until then, practice finding humor in failure. Walk up to at least one person you've never met, and just say hi. You may be surprised by how well it turns out.

KNOW YOURSELF

———

SOUTHWESTERN VIRGINIA

Bennett woke up the next morning with various Mikeisms circling through his head. The idea of chutzma was intriguing, and the story of a kid fake playing cello for a whole school year was hilarious. But most importantly, the podcast had left him with a question he hadn't thought about in a very long time.

Who am I?

Still mulling it over, he fed Maurice, strapped on the harness, and took him out for his morning walk. He briefly thought of making it a morning jog. What was it Mike Shapiro had said? *Find a way to make fitness a part of your daily experience. Whatever it takes to keep moving.* But then Bennett remembered how out of breath he'd gotten yesterday and decided to stick with walking. No one ever mistook a coder for an athlete, least of all him.

He had always liked his grandmother's neighborhood, ever since he was a kid. The houses were brimming with southern charm. Most homes had wide, sweeping verandas dotted with plants and flower-pots. You could tell people put a lot of thought into their landscaping, with nicely manicured lawns and late-spring flowers blooming in the flower beds. As Maurice sniffed around various front lawns, Bennett inhaled the clean, crisp smell of fresh-cut grass.

I'm someone who appreciates the little things in life, he thought. He looked at Maurice, noting how happy he looked in the morning sunshine, tongue lolling as he trotted along. *I'm someone who loves my dog.*

All *dogs,* he amended. Because the truth was, Bennett had never met a dog he didn't like.

I'm someone who likes making people happy, he thought, remembering Todd and the app. *It's why I first started making apps. I enjoy making things a little bit easier, a little bit brighter, a little bit better for other people.*

An epiphany fired up in his brain. He'd never thought about it this way before, but one of the things that brought him the most joy was actually bringing joy to other people.

Maurice stopped in front of an impressive two-story home with a wraparound porch that seemed to go on for miles. Bennett paused to admire the house himself. It stood grandly on the corner lot of Elm and Main—prime real estate. It was clearly expensive but tastefully designed and decorated, not at all ostentatious. He noted the stately white pillars on the front porch and a cluster of Adirondack chairs around a rustic redwood coffee table with, appropriately, a hot cup of coffee sitting on top. He could tell it was hot by the way the steam curled out of the mug.

Bennett also did not fail to notice the two vehicles in the driveway: high performance cars so new they still had temporary plates. He let out a low whistle. Whoever lived here had clearly struck gold.

That was all it took to plant a seed of doubt in Bennett's mind.

A moment before, he'd been thinking about the things that gave him joy and what might comprise his "inner genius." Now, staring at this nice home with these two *very* nice cars, he imagined the person who had brewed that coffee. Someone wealthy and successful. Someone who could spend a leisurely morning—on a weekday, no less—sipping a cup of black coffee on their picturesque front porch.

Meanwhile, there stood Bennett, unemployed, living with his parents. He didn't even own *a* car, let alone a nice one. He wanted so much more.

His cheeks went red hot with embarrassment. All the same old tapes began playing in his head. He'd been fired from a job that would have made him wealthy, all because he couldn't give it 110 percent. Why did he keep failing so abysmally? Would he ever have what it takes to succeed?

Bennett suddenly felt very small and very out of place in his grandmother's neighborhood, among all these people who had clearly made the right choices in their lives.

He saw movement inside the house. The knob of the front door began to turn. The coffee drinker who had made all the right choices was coming back outside.

Bennett tensed. He heard Mike's parting words in his head. *Walk up to at least one person you've never met and just say hi. You may be surprised by how well it turns out.*

The front door swung open.

But Bennett didn't wait to see who walked out. He was already walking swiftly in the opposite direction, cursing himself for his cowardice, and feeling smaller than ever.

What did I think? he thought angrily. *That one podcast episode would really change my life?*

. . .

By the time Bennett made it back to his grandmother's house, he was in a lousy mood. It didn't help that he'd spent the whole walk back making a list in the Notes app of his phone of all the things he was not.

I'm not …

- a risk-taker
- brave
- talented
- hirable
- someone who can give 110 percent or even 100 percent

No wonder his bad mood spread faster than a prairie fire with a tailwind. He clearly wasn't doing the what-you're-not list right. Surely the point of the exercise was to help him narrow down what he *enjoyed* doing, not wallow in self-pity.

As he trudged through the kitchen, his grandmother greeted him with a smile—and the heavenly aroma of blueberry pancakes.

"Hungry?" she asked. "I know they're your favorite."

"They smell delicious. Maybe later," he said, not wanting to hurt her feelings. But he wasn't in the mood for pancakes.

She shrugged. "Suit yourself. I'll put a plate in the fridge that you can heat up later."

He nodded, hoping he hadn't seemed rude. "Hey, Gran, who lives at the corner of Elm and Main? It's such a great house."

She cocked her head. "Are you pulling my leg?"

He blinked. When she saw his earnest confusion, she laughed.

"What's so funny, Gran?"

"Why, Ben, don't you know? That's your Uncle Jim's house!"

Bennett was staggered. The last time he'd visited, his Uncle Jim and Aunt Emma lived in a very humble one-story home in an entirely different neighborhood across the train tracks. More of a cottage, really.

Now he conjured up the house on Elm and Main, remembering the luxury cars in the driveway.

"But ..." he stammered. "I saw Jim yesterday. He had a brand-new F-150!"

"He sure does. But Riley's home from college this week for spring break, and the Aston Martin ... don't even get me started. When Jim came home with that truck, Emma was furious. Said she wanted something fun to drive herself! So he got her that Aston Martin for their anniversary."

Bennett couldn't believe it. Since when did Uncle Jim have the cash to buy his wife an Aston Martin? What in the hell was Jim *cooking* in that restaurant?

"You should go say hi," Gran suggested. "He's probably up at Mill Creek now. He makes it over there most mornings, just to chat to customers. That youngest boy of mine always did have the gift of gab."

Which was how Bennett found himself making the short trip to Mill Creek Restaurant. It wasn't until he pulled into the parking lot that he realized how badly he needed to talk to someone, someone he trusted.

And there was Uncle Jim, standing out front, waiting. Bennett had a hunch that a particular southern grandmother might have tipped Jim off that his nephew was en route.

"Heard you came by the house this morning," said Jim with a sly grin, as Bennett jumped out of his truck. "Why didn't you say hi?"

"I had no idea that was your house," Bennett said. He did a double take at the restaurant's exterior, which looked nothing like he remembered. Where was the humble little country food dive? The elegant establishment behind his uncle looked worthy of a Michelin star. "To be honest, I wouldn't have known this was your restaurant either."

Jim chuckled. "A lot of things have changed around here since the last time you visited. Come on in, son, and I'll show you everything you missed."

• • •

Bennett's first impression of Mill Creek Restaurant was that it smelled the same way it always had. The air was spiced with the aromatic scent of ham frying, biscuits in the oven, and piping hot coffee. He even recognized a few familiar faces in the hostess and servers, people who'd worked with his uncle for years.

But that was where the familiarity ended. Because the restaurant looked … nice.

Uncle Jim's restaurant had always had that "homey diner" feel, with its tread-worn vinyl flooring, frayed pleather booths, and fake flower arrangements. That was all gone now. The flooring was real wood, varnished to a fine sheen, and the booths were reupholstered in soft, sturdy suede with nary a dusty fake flower in sight. Each table had a thin glass flute with a single long-stemmed red rose.

"That's my table," said Uncle Jim, pointing to a cozy corner booth. "Go grab a seat, and I'll pour you some coffee. You want breakfast?"

"Gran made blueberry hotcakes, but I passed. I'm not really hungry."

"You turned down Mom's famous hotcakes?" Jim made a big show of pressing a hand to Bennett's nonfeverish forehead. "You sure you're feeling all right? You know, I've been trying for years to get her to give me the recipe. I own a restaurant, and she still won't budge. She's like Fort Knox; you can't get *anything* out of that woman. She guards her recipes like state secrets! But come on, at least let me bring you some biscuits and gravy. You know ours are the best."

Bennett couldn't argue with that and didn't plan to. Jim disappeared behind the swinging kitchen doors as Bennett made for the corner booth. Neat piles of paper were stacked on the table. Before he could get a good look at them, his uncle reappeared with a plate of hot biscuits, a ceramic gravy boat, and a cup of coffee. He set everything on the table.

Bennett took a sip of coffee, then another. Even the coffee was different. This was no longer the tea-pale diner coffee Jim used to serve. It was dark and rich, full of complex notes and a well-rounded finish.

"This coffee is amazing," Bennett said, as Jim slid into the booth beside him. "And the biscuits are as good as they ever were. But that's about the only thing I'd recognize from before. The whole place looks amazing, Uncle Jim."

"It's a makeover this old place has needed for years. Just a few touch-ups here and there."

"More of a major renovation, I'd say."

"It's been a fun project to reenvision Mill Creek and what it could be. We got write-ups in a couple of big papers. They called it 'the best diner food you've ever had, with a modern upscale twist.' Or my personal favorite: 'comfort food with class.' That's when people really started pouring in. Did you see the ceiling fans?"

Bennett glanced up to see four sets of wide, slowly spinning blades scattered around the dining area.

"I've never had so many fans before," Jim said, chuckling at his own joke as he combined the stacks of papers, clearing more table space—but not before the words "Real Estate Purchase Agreement" caught Bennett's eye.

"Now that we have the fans," Jim said, "I don't know what we ever did without them. You know how it gets hot as blazes around here in summertime. Even in spring, when you're slammed on a shift and in the weeds, you really start to sweat."

He caught the question in Bennett's eyes and followed his gaze to the papers.

"Yeah, I've had a lot going on since the last time you visited. This," he said, gesturing toward the paper stack, "is the most recent project. Do you remember Sophie Miller, the fabric shop owner? Well, as soon as I get this sorted, I'll own the building her store is in, as well as the one next to it. Which puts me at … let's see …" He leaned forward, peering out the window, eyes flicking up and down the brick-store-lined main street. "Nine buildings, total. By this time next year, I plan to own at least half of the commercial buildings in this three-block section of town."

Bennett stared at his uncle, wide-eyed. The last he remembered, Uncle Jim was barely making ends meet with his restaurant.

"Did you win the lottery?"

Jim laughed so hard his face turned red. He slapped the table.

"I wish!" he said, wiping his eyes. "This would have been so much easier if I did. But no, this is just the result of hard work and finally learning who I am and what I really care about."

It started, Jim explained—settling back in a way that Bennett knew he was in for a long story—a few years ago with Joe, the owner of the barbershop a couple of buildings up from Jim's restaurant. Bennett leaned back into the booth contentedly as his uncle began to spin his tale.

• • •

It had started like any other visit: Jim came into the barbershop for his regular monthly trim. But instead of diving into yet another wildly elaborate story as he sat down, something about Joe's expression made Jim hold back.

"Lumberjack chop off that chatty tongue of yours, Joe?" Jim asked jovially.

Joe glanced up in the age-speckled mirror and managed a slight grin. "Ah, sorry about that, Jim. Today's not a good day."

"Why's that?"

"Well, I think I'm going to have to close."

Joe looked down at the scissors in his hand, wiped them absently on his apron, then glanced up at Jim in the mirror again.

"The property manager dropped by about an hour ago to say the owners of this building are jacking the rent up almost double. I can't afford that, Jim. I barely make it work as it is. But I love this place. I've been here more than forty years if you count the time I spent here as a kid watching my dad work, rest his soul."

Jim nodded, not sure what to say, apart from "Sorry to hear that."

Joe, quiet, went back to trimming his hair, the sounds of the local radio station barely filling the silence.

Jim took it for as long as he could, then, "You know what your problem is, Joe? You don't look at the bright side enough. Just think of what you could do if you didn't have to trim hair for schlemiels like me every day. Go fishing any time you want, sleep till noon, write a novel, get a camper and explore the country, find yourself. The world is your oyster, my friend. Why not go pry it open?"

Joe smiled and slowly shook his head.

"I'm doing what I love, Jim. This shop. The people like you who come by. I know everyone in this town. Been cutting hair for most of them since they were old enough to sit up on their mother's lap. I don't need to go 'find myself'; I already know who I am—a barber, in this town, in this shop. And I don't know what I'll do without it."

On his way out, Jim asked, "How long do you have? Until the rent goes up, I mean."

Joe scratched his thinning hair. "About six months, I think? It should be enough time to get a plan in place, I hope."

Jim tipped Joe a little extra, even though he didn't have a lot of extra cash himself those days, and walked back to the restaurant with a lot more on his mind than when he left.

> Success follows happiness, not the other way around.

A few weeks later, still thinking about Joe's tough spot, Jim was walking through the kitchen and overheard one of his cooks listening to a radio program while prepping for the dinner service. Normally he didn't give a second thought to what they listened to—whatever kept the team happy while they chopped vegetables and peeled potatoes—but this time something he heard made him stop in his tracks.

"Developing a deep belief in yourself and your capabilities, strengthening that belief, making yourself stronger through that belief, and getting others excited and engaged in you and your cause …"

Jim lingered in the doorway for a minute, listening. The speaker was explaining the difference between working hard and enjoying your hard work—how success follows happiness, not the other way around.

It made sense.

Not only that, it made something click in his mind regarding Joe's problem and his own feelings of being stuck at the restaurant, with no real sense of direction or purpose. Because, the truth was, he hadn't been excited about Mill Creek in years. He'd sat there and schooled Joe on how the world was his oyster, but when was the last time he, Jim, had pried it open?

"What is that?" he asked from the doorway, startling the cook so much that he almost knocked over a full pan of gravy.

"Sir?"

"The radio. What are you listening to?"

"It's not the radio; it's a podcast," the cook said, picking up his phone and pointing to the Listening Now screen. On the icon below, Jim could make out three little words: *Read the Tape.*

Jim nodded; threw in a quick "Ah, right, thank you"; and walked away, making a mental note to google *how to listen to podcasts* as soon as he got home.

• • •

A month later, Jim had listened to the whole series twice. Not only did he know more about himself than he ever had; he also knew something monumentally valuable: what made him happy. And part of that was realizing how much joy he found in making *others* happy. He'd found his inner genius.

That day, he dug into his meager savings account and talked to his banker about a second mortgage on the restaurant. That night, he had a heart-to-heart with his wife, Emma, where he laid it all on the line. With her blessing and support, he gathered up his courage the next morning and booked a meeting with the out-of-town owner of

51

Joe's commercial building. A few weeks after that, he was the proud owner of a second building in the main shopping district of town.

He was also dead broke.

Sitting in his basement office, going over the stack of bills, Jim felt sick to his stomach. Even if he doubled Joe's rent, he still couldn't cover the costs on both his building and the barbershop—and he wasn't going to do that to Joe. What would Emma say? Their oldest daughter, Marcy, was a few years out of college, but Riley was only a sophomore, and it was already putting a huge financial strain on them to pay her in-state tuition. What the hell was he thinking?

Still, something inside him wouldn't let go. There just had to be a way.

The old wooden chair in his basement office creaked as Jim leaned back and took a long, deep breath.

He glanced at the clock: 8:30 p.m. It wasn't too late, and he needed some air. And nothing sounded more appealing at that moment than a walk down to Tilly's.

Tilly's was on the other end of the "downtown district," which meant it was about a ten-minute walk away. Its Gothic, hand-carved wooden door had been installed sometime in the 1920s, and the warmly glowing, crosshatched stained glass windows gave it the feel of old Ireland—which it should have, since it had been family owned by the Callahan family since the town was settled in the late 1800s.

Jim ambled up to the worn wooden bar and slid onto one of the comfortable and equally worn stools. Before he could tip his imaginary hat at the bartender, she already had a pint in front of him.

"You never forget anyone, do you, Sam?" He smiled at the bartender. Samantha laughed.

"Now, how can anyone forget you, Jim? I swear, everyone in this bar right now knows who you are … and I'll bet you know every one of them too."

Jim surveyed the dimly lit tavern. She was right. There were Josh and Nate from the repair shop in earnest conversation at a window booth, Sophie from the fabric store chatting it up with Elaine from the radio station, Mr. and Mrs. Strickland from the grocery, and the whole ten-person crew from the town's rec center. A few of them saw him glance their way and raised a glass. He did the same, then turned back to Sam.

"You know, Sam, as nice as it is knowing your neighbors, sometimes I miss the joy of meeting new people in town. They're just so few and far between these days."

Sam was nodding politely when something behind Jim caught her eye, and she glanced up with a slight look of surprise. "Well, now's your chance," she half whispered to him, then to his left. "Hey there! What can I do ya for?"

A couple of stools down from Jim, a young man he'd never seen in town before hung his jacket on a bar hook and tucked into a spot.

"Something local, if you have it. Maybe an IPA?"

She nodded and raised an eyebrow at Jim as she headed to the draft wall.

Jim didn't hesitate.

"My friend, if you like IPAs, you're really going to like the one she's pouring you. Two young brewmasters just down the road make it, and it's my go-to."

The young man smiled as Sam slid the pint in front of him, and he held up the glass. "Cheers to that," he said. Jim did the same.

"So what brings you here?" asked Jim as Sam walked away again, laughing softly to herself and shaking her head. That Jim.

In less than fifteen minutes, Jim knew just about all there was to know about Ethan Grinstead—where he was from (Ohio), what brought him to town (traveling for business), and the fact that he was an investment banker, which piqued Jim's interest considerably.

"So let me ask you a question, Ethan," said Jim, gesturing to Sam that the next round would be on him. "I have a couple of commercial properties a few blocks down on Main, and one has two floors of open space with a barbershop on the ground floor, but I can't get anyone interested in renting them. What would *you* do?"

Ethan rubbed his road-weary five-o'clock shadow in thought, then offered, "Well, there are a lot of factors that go into it. What's the space like? What's the foot traffic? Is it zoned for residential, too, or just commercial?"

As Ethan listed off questions, Jim walked him through the whole setup, from location to appearances to zoning, and yes, it was residential-commercial and the spaces were impressive—hundred-year-old oak flooring, brick walls, open floor plan. If these were in a larger city, they'd go for top dollar. But Ethan brought up a very, very good point. Even on a weekend, there was very little foot traffic in town. It was an attractive, old Main Street, but the few shops were tired and dated, and most people just parked in front of whichever antique store they were visiting and left. There was nothing interesting enough to encourage a stroll or window-shopping.

"Well," said Ethan, "what if you made it interesting? Tell me, does your town throw a lot of festivals?"

The realization hit Jim like a confetti-filled brick. *Why in the world didn't I think of that?*

"Why in the world didn't I think of that?" he said out loud. "Festivals! I mean, sure," he said, counting them out on his hand.

"There's the New Year parade and the Fourth of July parade, the May Day festival, and … Sam, help me out, what's the other thing we do?"

"Apple festival," she called out from behind the tap.

"Apple festival. Right. But that's about it."

"So why not fill those empty weekends?" Ethan said eagerly. "Bring in vendors, set up tents, do some food trucks, maybe even have a band play. Give people a reason to start walking these very walkable streets, and they'll start looking into how they can be around here more often."

"Son," said Jim, hopping his heavy frame off the stool with all the energy of a teenager, "I owe you more than you'll ever know. Sam, his tab is on me." He glanced quickly at his old Timex watch. "It's not too late for a few phone calls. I gotta run. Ethan, do you have a card?"

Ethan handed him one—expensive looking and embossed on thick cardstock—which Jim quickly pocketed.

"Perfect. Now I know how to reach you so I can return the favor. Safe travels, and if you're ever in town again, look me up. I'll owe you dinner and a beer at the very least."

• • •

Jim did more than call a few people that night. The next morning, he was at the town offices filling out festival permits, and by noon he was bartering with the owner of the rental company on bulk tent rates. By early evening he had a band lined up and several business owners, from the shopkeepers to local craftspeople, had agreed to rent a tent for the event. And for each one he got on board, he asked them to tell as many of their friends as they could. Market on Main wasn't going to be a onetime deal; it was going to be a regular driver for new business, new residents, and tourism for as long as he could keep it going.

It was at some point in the middle of all the planning and phone calls and negotiating that Jim suddenly realized he'd never felt so alive. He'd put everything he had out on a limb—practically a twig, if he was honest with himself—and by all accounts he was terrifyingly deep in debt. Yet he'd never been happier. Sure, he could fail at any moment. He already had enough naysayers telling him he was investing way too much in properties, and now this, and that he should stop before he lost everything. Even Emma was skeptical. What if he lost all their savings and Riley had to drop out of college?

But deep down Jim believed not just in the event but in the people. And because of them, he knew this would work. It was a gut feeling, sure, but he'd learned that oftentimes that's exactly where you need to put your trust.

Despite his confidence, on the day of the first Market on Main, Jim was a nervous wreck. Every fear came bubbling to the surface: What if it rained? What if no one showed up? What if everyone canceled? What if they hated it? What if Emma left him and Riley never forgave him and everything went to hell in a handbasket?

It wasn't until the last shop finished setting up and the first attendees began to trickle in that his confidence returned. And as the crowds poured in and the strumming of the band and the mouthwatering smell of home cooking drifted on the air, he allowed himself to breathe again.

All told, the first Market on Main wasn't an overwhelming success, but it did well enough that Jim was able to meet his most immediate financial obligations and set some aside for the next event. Things grew quickly from there. Every month on the first Saturday, rain or shine, Market on Main would take place. Even as Jim was helping fold up tents and pack out trash that evening, he was getting calls from out-of-town vendors and—to his surprise—even a couple queries about his rental spaces. It was all coming together, and he already had big plans for what it could be.

Jim felt something new that night, something he'd almost forgotten the feel of. *Joy.* In spite of all the anxiety and risk and uncertainty, Jim felt pure joy at what he'd created.

He was following his happiness to success.

• • •

"And here we are," said Jim, rapping his knuckles on the restaurant table and jolting Bennett back to the present. "I jabbered on so long we might as well order lunch!"

As usual, Bennett had lost all track of time. It was always like that when his uncle talked; he was completely swept up in the story, right there with Uncle Jim every step of the way.

"After the first three Market on Main events," Jim said, "I hired my prep cook full time to run the social media and coordinate vendors for me."

"Your prep cook?" Bennett said, puzzled.

"Yeah, the one I heard listening to the podcast. The kid's pretty smart, and it turns out he has more artistic talent in his big toe than I have in this formidable figure combined." Uncle Jim gave his stomach an emphatic pat. "I asked him what he thought of it, and he ended up telling me all about his marketing company start-up and how he's working odd jobs like prep cooking just to pay the bills. So I asked him to help me out with some marketing for the first event and gave him a small budget, and he nailed it. I was hearing about Market on Main on the radio, seeing it on posters. Even overheard people talking about it at the grocery store. He also had this great idea for a vendor-tent giveaway that went bonkers. Short story long, he's killing it. Now, not only does he do the marketing, he also books the craft tents, hires the musicians, and coordinates the food trucks. All I have to do is what I love doing anyway—make connections and talk everyone's ears off."

Bennett smiled. His uncle never ran out of things to say. Gran wasn't kidding; he really did have the gift of gab. When Bennett was a kid, he used to wonder what Uncle Jim would do if he didn't have anyone to talk to. He'd heard Emma joke that if something ever happened to her, Jim would be just fine since he could probably coax a wall or a rock to start a conversation with him.

"All that from a podcast, huh?" Bennett said.

Jim shot him a wry grin.

"It wasn't the podcast, Ben. The podcast didn't give me the idea or even tell me what to do. What it *did* was get me thinking. For a while, I'd been stuck in this fool notion that I was too old to make any big changes in my life. I'm not a young buck like you! I've got two grown daughters—surely that meant I was too old to be an entrepreneur. But then Mike said something I'll never forget: "Age doesn't matter." Nobody is ever too old. It's always a good time to start something, to start over, to begin, to achieve. There's never a time *not* to do that. It's just about finding your passion and the things you love most. That's the important question.

"That reminded me that I've only got this one shot at life. If I'm not enjoying myself, then what am I doing? If I don't give myself permission to fail, how will I ever know if I can succeed? What I'm learning is that when someone, anyone, is given the chance to do something that means a lot to them—something that gives them a sense of purpose—they're almost always going to pull it off. They just need the courage to try."

Bennett looked down at his coffee mug, a mix of emotions.

Just try. That was all well and good for his uncle. He was an experienced businessman long before he came up with the Market on Main idea. He had resources to draw on, connections to tap. What did Bennett have? A dog who thought he was ten times bigger than he was, the bag of clothes he brought with him, and maybe a few new Note files on his phone.

Just try.

With what? How?

"Don't think too hard; you might break something," Jim said, busting Bennett's reverie. "Tell you what. Tomorrow's the first Saturday of the month, which means we've got our next Market on Main all cued up. Why don't you come down tomorrow and see for yourself? We got a band driving all the way up from Nashville, and two of the food trucks are run by James Beard award winners."

"Sure, Uncle Jim." Bennett nodded. "I'd love to."

"I know you're more of a techie guy, but that's all the more reason you should come out tomorrow. We've got a couple of new booths this month that we're trying out. I got approached by a Silicon Valley type who wanted to put on a little mini Maker Camp for the kiddos with some cool sound machines and lasers. I said, Why not? The kids can amuse themselves while their parents listen to good music and chow down on great food. Tell you the truth, I was pleased as pie. Never thought we'd garner interest from big tech with our humble little street fair."

Bennett tried not to smirk. Lasers and sound machines? Funny what passed for "big tech" in southwestern Virginia.

"That sounds fun, Uncle Jim. I don't have a kid to bring, but how about Maurice?"

Jim guffawed. "You're still young, Ben. You got plenty of time for kids. And until then, all dogs are welcome at Market on Main."

"Great. We'll be there."

Jim stood and gave him a hard pat on the back. "You're gonna love it. When you get there, just come find me in the tent by the bandstand. And in the meantime, keep listening to *Read the Tape*. Take it from me—it just might change your life."

• • •

Welcome to the next episode of *Read the Tape*. I'm your host Mike Shapiro, and here's one thing I know about myself: I'm confident. Are people born with confidence? Well, yes. *I* was. All kidding aside, confidence is built through successes. Every time you have a success, you get more confident. Yes, you're going to fail along the way. But it depends on what you do with those failures. Turn it into something that's successful so that each time you're successful, more confidence comes. Then you're *always* confident.

Here's something else I know about myself: I have a great sense of humor. Laughing? That's everything. If you're not laughing, you're not living. I laugh all the time. What else are you supposed to do? *Cry*? Keep laughing. It's the most important aspect of your daily experience. Staying in a good mood, being positive, enjoying life. Laugh at everything!

I promised you that in today's episode we'd talk about knowing yourself. I didn't lie. But part of knowing yourself is *enjoying* yourself. I hope you've been thinking about who you are, what brings you joy and makes you happy. And I also hope you did your homework and wrote down five things you're not. If those five things made you feel like a failure, guess what: you're going to fail along the way. There will be failures. Your success depends on what you *do* with those failures.

Bennett shifted in the driver's seat. It was a little uncanny how this Mike character always seemed to be talking directly to him, Bennett Gates. Because of course, that was exactly what he'd done, used the what-you're-not exercise to feel like the biggest failure of all time.

As Bennett drove away, he mulled over what Uncle Jim had told him. The whole thing sounded like a massive risk on Jim's part, and yet, from the looks of his restaurant and the growing number of properties under his belt—not to mention the gorgeous new house and cars—he was pulling it off.

Not just pulling it off, Bennett corrected himself. Jim was actually becoming really, really successful at it. He had never seen his uncle happier.

He turned the volume up on the podcast, hungry to hear what Mike would say next.

We get better at the things we like to do because we enjoy doing them. Sounds obvious, right? You're probably thinking, "Tell me something I *don't* know, Mike." But what most of us don't realize is that this understanding is foundational to success.

Succeeding isn't just about money or notoriety. It's about doing something that gives you purpose. It's about truly enjoying yourself—about *being* yourself and embracing your inner genius—and finding that, in the process, you're becoming really good at what you do.

It's also something you have to decide for yourself. No one can tell you what your inner genius is because no one knows *you* better than yourself. And ultimately, *no one has the power to change your life circumstances as much as you do.*

> No one can tell you what your inner genius is because no one knows you better than yourself.

Bennett found himself nodding along. In practice, he hadn't felt particularly powerful recently, but in theory it made sense that he knew himself better than anyone, which gave him the best shot at unlocking his inner genius.

When I was a kid, I naturally gravitated toward doing those things that felt "right" to me. The cello incident, which I talked about in the last episode, was a natural solution to meeting extracurricular requirements while avoiding the destruction of a perfectly good orchestra. Plus, it let me exercise my budding acting skills, and my dad got a real kick out of it.

Around that same time—I think I was about eight or nine—I also learned that I really liked helping other people, especially those that couldn't stand up for themselves.

Now, in school, I was never a small kid. Maybe it was the swimming classes that bulked me up, or maybe I actually had some kind of muscle condition, but regardless, I was very muscular for a nine-year-old. You'd think that would be a good thing—who doesn't want to be ripped?—but what actually happened was that everyone, from the bullies to the teachers, body-shamed me. To this day, I still recall one of my teachers yelling at me over and over again to "Put your arms down!" But I couldn't. They just puffed out like that poor snowsuited kid in *A Christmas Story*. I went home that day bawling because my arms wouldn't go down. But instead of being angry at the teacher, my dad thought it was hilarious.

"You won't believe me now, but one day, you'll see this as a good thing," he laughed. "For now, just let it go." And he was right. The other kids made fun of me. So what? That didn't make them better than me. So the next day, I just rebuffed their teasing by accepting that, yeah, my arms are big. So what? With no emotional response to keep them going, they eventually just let it go.

To my dad's point, I didn't have to wait long to see a benefit to having overdeveloped muscles. By the fourth grade, I was what you'd call a "bully's bully." If someone was being picked on or hurt because they were small or weak or smart or just different, I stepped in and stopped it. I wouldn't hurt the other person; just the fact that I was big was enough to make them walk away.

And that's exactly what I did for this girl who was being picked on, not only by the bullies but by pretty much everyone in our class, mainly because she wouldn't stop picking her nose.

After fending off the worst of them for what felt like the hundredth time, I finally turned to her and said, "I don't mean to hurt your feelings, but why do you do that? You know the other kids would stop picking on you if you stopped picking your nose."

Just like that.

It was a weird thing for a ten-year-old to do: confront a peer so directly about an issue that was clearly causing her problems. But it was such a simple fix that I didn't even think about it. I just put it bluntly. And you know what? It helped! She didn't even realize she was doing it, and once she did, she was thankful that I pointed it out.

Her happiness and seeing how it really helped was the first time I realized how much I liked helping people. Six years later, I learned how that passion played into another natural trait that brought me joy: leadership.

By now, Bennett had made it back to his grandmother's house. But he was hooked on the story and didn't want to stop listening. He decided to stall by circling the block, biding his time.

I was sixteen years old when I got my first job as a "burger-assembly artist" at a well-known fast-food chain. It was only a few hours after school, but I made the most of it. I became known as the guy who always shot burgers into the heat-rack slide like a bartender slinging a fresh pint.

I was probably a terrible employee, but that's not the point. What was important about that job was that a few days into it, I realized that there was a great injustice being done to me and my fellow employees: by franchise rule, all employees had to work *four hours* to get a free burger. Yet, since this was a part-time job for me, I was never scheduled to work more than three. My fellow after-school employees and I worked our tails off during those hours. We deserved a burger!

I was so convinced that this was a fair cause to fight for that I led a walkout strike.

We were so successful that the chain location was forced to close for the day, and the protest even made it to the local news. I remember feeling incredibly proud of myself and was about to tell my family all about it when the phone rang. It was the manager, some kid barely older than me, screaming about what I'd done and demanding that I bring my uniform back.

Once again, my dad thought it was hilarious and told all his friends about his son, the labor leader.

And yes, I know. I was incredibly fortunate to have such a strong support system in my parents. Most folks don't get that kind of constant, positive encouragement, and what that meant to me was that I felt easily capable of both leading and not worrying about the consequences of pursuing what I believed to be right.

Aside from wonderful parents, however, there are several things you *do* have control over that will make all the difference between success and failure. We talked about one of these traits earlier: chutzma, the combination of self-confidence, or chutzpah, and charisma.

Self-confidence comes with knowing who you are and being okay when things don't go your way. It's about believing in yourself to the point that you're willing to take risks and, if things don't work out the way you hoped, letting the results be what they are and moving on.

But for self-confidence to really settle in, you also need to build up two more all-too-important traits: resilience and a sense of humor.

When I mentioned swimming earlier as a reason for my too-young musculature, it was because swimming was my whole world back then, and being on the swim team took a lot of all these traits—resilience, self-confidence, and more than just your run-of-the-mill sense of humor.

There was one meet in particular that I'll never forget.

We were the visiting team at a school up north, and I was prepping for one of the longer back-crawl events. Along with the other swimmers, I walked the edge of the pool, rolling my neck and shaking my big arms and legs to increase circulation and loosen up the muscles. Then I hopped in and resurfaced in a race stance … only to hear peals of laughter echoing around the cavernous hall.

I was on the wrong end of the pool.

Some kids would have quit in embarrassment. But just like my parents who were in the stands, laughing so hard that tears rolled down their cheeks, I thought it was hilarious.

When you're comfortable enough to laugh at yourself, you also find that a good sense of humor gives you a certain degree of fearlessness. Why not just try when the worst that could happen is a good laugh? This sense of resilience has helped me to shake off setbacks and see failure for what it is, a chance to learn what doesn't work and focus on what will.

As for that swim meet, it probably won't shock you that I didn't win. But I did get better, and that unwavering sense of self-confidence and dismissal of failure eventually led to becoming a state champion.

I knew I could. I just had to give myself the space and time to get there.

This leads to the last trait of success that you have the ability to control—and it's a big one.

What I'm not.

We talked about getting to know yourself in the last episode, and now we're going to take that a little deeper. Remember the homework assignment I gave you, to write down five things you're not and one thing that brings you joy? If you have it, grab it. If you don't, feel free to pause this whatever it is you're listening on and jot some down.

The timing worked out well. After circling the block several times, Bennett had just pulled back up to his grandmother's house. He saw Maurice's eager face in the front window, his tall, tan ears pricked at the purr of the truck's engine. Maurice always seemed to know when his human was coming home.

Bennett reached for his phone and opened his Note file from earlier.

I'm not ...

- a risk-taker
- brave
- talented
- hirable
- someone who can give 110 percent or even 100 percent

He had to chuckle, a dark chuckle. Surely not even Mike Shapiro could turn *that* list around.

Lately I've noticed a pattern come up for a lot of the people I coach. These are talented, smart people, and they get so close to the thing they want—and then they pull back. I call them ninety percenters, because they get 90 percent of the way there. I sit there thinking, *If you get 90 percent of the way to the wall, why don't you touch the wall?* But of course, if they take that last leap, there

> These are talented, smart people, and they get so close to the thing they want— and then they pull back. I call them ninety percenters.

could either be absolute failure—or the great success they want. Right now, they're coasting, and as long as they're coasting, they can keep the failure at bay. So, sure, they have a fear of failure. But they also have a fear of success. They fear success because then they'd have to act on it to keep going.

Bennett was glad he'd pulled over, because if he'd still been driving, he might have crashed the car.

A ninety percenter. His mouth went dry, his heart pounding in his chest. That was *exactly* what he was—in jobs, in relationships. He got 90 percent of the way to the thing he wanted … and then didn't touch the wall.

He was almost afraid to keep listening to the podcast. But the pull was too strong. Maybe, if Mike really did help other ninety percenters, he could help Bennett too.

Now, we're doing this because figuring out what we're *not* jolts us out of our natural-response process— that is, most of us have an automatic reply when asked "What do you enjoy doing?" It's pretty much expected as part of any job-interview process, and it's something we're used to sharing in conversation.

But when we look at what we're *not*, we get a much better sense of who we *are* because we have to stop and really think about the answer. And it opens our eyes to things that we may never have realized were part of our nature. The goal of this exercise is to get you to see that, sometimes, being honest about what you're *not* can help you start to get excited about what you *are*.

If I say "I'm not comfortable being part of the crowd," for instance, I learn a lot more about myself than if I say "I have strong leadership skills." Now I'm looking at the negative space of what I'm *not* rather than stating something that most people find obvious. If I'm *not* one to blend in with everyone else, what does that make me? That may mean I feel the need to stand out or that I don't settle for average. It makes me think about my ability to play well with others and consider that I have a natural reaction to go against the grain. I could go even deeper and think about how this also lends to my habit of stepping back from the crowd and observing the room—what I call "Reading the Tape," which I'll dive into in our next episode.

But enough about me. What did you put on your list?
What do your "nots" tell you about who you are?

Bennett's heartbeat was no longer racing quite as fast as it had been. He squinted hard at his list. Sure, he wasn't a risk-taker. But he'd always been more of a planner. Ever since he was a kid, he didn't do things without really thinking through them first. He was someone who liked to take his time, plan things out, and do them right.

Which, when you put it like that, didn't sound so bad. Wasn't that the whole reason he liked developing apps? It was all about thought, planning, and execution.

Next up: he wasn't brave. Was that true? He thought about it. What he'd really meant by that, he realized, was that he was shy. He'd always felt shy about meeting new people—which was why he turned tail at Uncle Jim's house earlier and also why he hadn't yet taken Mike's advice and introduced himself to someone new. But surely that was a skill he could work on.

As for not being talented? Just because he wasn't some kind of tech celebrity didn't mean he didn't have talent. The truth was, he had a gift for writing clean, solid app programs. He always had. He just hadn't gotten a chance to flex his true creativity for years.

Bennett felt a little jolt of insight. He hadn't been given the opportunity to feel talented, because he'd been working in a job that made him feel burnt out and ground down.

As for not being hirable?

The truth was, he hadn't really *wanted* to be hired at any of the jobs he applied for. He didn't want to go back to the soul-draining, open-floor-plan, pseudocommunity popularity contest that was his long-standing impression of big tech. He wanted something more, something real, something that would make a difference and that he—ultimately—would drive.

He suddenly felt excited. Maybe the reason he didn't want to be hired at any of the jobs he'd applied to was because there was something else for him out there. Something he could put his whole heart and talent and passion into.

He just had to figure out what it was.

And last but least: Okay, fine. At his last couple jobs, he hadn't given 100 percent. He was absolutely a ninety percenter. That was something he needed to work on.

But then a new thought occurred to him. At these last couple jobs, wasn't it also true that he hadn't really felt like the work he did mattered? He was coasting, yes. He was putting in 90 percent, absolutely. But he'd also set himself up to fail. At the company with his college buds *and* at the video game start-up, he definitely wasn't having an objectively positive impact on other people's lives. And wasn't that what brought Bennett joy, making people happy?

Like Uncle Jim, he thought. Jim had a very different way of connecting with people—he said a hundred words for every two of Ben's—but he, too, liked making people happy. Bennett enjoyed creating more time and space for others to do the things that brought *them* joy, whether that meant presenting a bespoke app to his cousin Todd or throwing a ball for Maurice (which technically meant he liked making people and *dogs* happy, but why draw a distinction?).

What Mike said next was right on cue.

> That brings us to your next homework assignment. Now that you can see where you're standing, what are you walking toward? What makes sense to you?
>
> What brings you joy?

I'll leave you today with that thought and ... wait for it ... some more homework. I know you were on the edge of your seat wondering what I was going to throw at you this time. And again, it's a little coun-terintuitive—kind of like the "not" list.

I want you to do something that makes you uncom fortable, something disruptive. Not something illegal, of course, but something you've been afraid to do even though you know, deep down, that it's really not that big of a deal.

Remember, in the last episode, we talked about that bodybuilder who was too afraid to talk to his idol? Go do that. If you're afraid to talk to someone you admire, or who's higher up on the chain at work, or who you just haven't had the courage to strike up a conversation with, go do it! Go talk to a random person at a bar, or say hello to someone in line at the coffee shop. Don't let fear or arrogance get in the way. Be open to the conversation, and you may be surprised where it takes you.

> If you're afraid to talk to someone you admire ... someone who you just haven't had the courage to strike up a conversation with, go do it!

I look forward to hearing how it goes. Until next time, keep reading the tape.

READ THE TAPE

SOUTHWESTERN VIRGINIA

It was late when Bennett finally made it out of the house the next morning.

After pouring a mugful of coffee and planting a kiss on his Gran's cheek, he and Maurice hopped in the pickup and took off once again toward the brick-lined Main Street where his uncle had promised an event to remember.

Bennett was still three blocks away when traffic came to a grinding halt. Every spot along the sides of the quiet, residential road leading into town was full, bumper to bumper with cars, many of them from out of state. He saw plates from Tennessee, Georgia, both the Carolinas, Texas, and even one from California. As he crawled along, he noticed the small groups heading toward Main were getting larger and ranged in age from infants to seniors.

By the time he found a spot and hiked the six blocks to Main, it was nearly noon, and the Market was in full swing. Breezy, white clouds, scented with bacon grease and sugary fried dough, drifted over the sea of shop tents and the happy, bustling crowd. Over the dull roar of chatter, Bennett could make out the twanging sounds of a band warming up and playing a few riffs of a country song. He looked up and was shocked to see not the pallet-and-plywood bandstand he'd imagined but a professional mobile concert stage, complete with lights, sound system, and a curtained backstage. He thought he recognized the song too.

Suddenly he realized exactly who the band was. This wasn't some bar-playing gig group; it was the real deal. The concert hadn't even started yet, and already the crowd surrounding them was a couple hundred people strong—and growing.

"Hey, Ben!" called a familiar voice. Uncle Jim stood in front of the large canvas tent behind the stage, beckoning his nephew over. Bennett hurried over, Maurice trotting along beside him.

"This is amazing, Uncle Jim. I can't believe it."

"Wait till you see the inside."

Bennett ducked through the canvas tent flap … and found himself in just about the best glamping setup he'd ever seen. He took in the heavy rugs, leather furniture, wide-screen TV, and full bar along the wall.

Jim's smile was wide and proud.

"Great, isn't it? I set this up for the band, on the condition I can use it while they're onstage and after the Market closes to host the after-party."

Bennett slowly shook his head in awe.

"I had no idea, Uncle Jim. And onstage, is that really …?"

"Yup, in the flesh! They reached out to us six months ago about performing. They're headlining today's show, but to tell you the truth, they're not even the hottest ticket we have this year. I tell you, Bennett, when you really decide what you want to do and you put everything you have into making it happen, chances are, you will. And if you don't, hey, it's better than never trying at all."

Jim dropped into one of the wide, leather club chairs and gestured for Bennett to take the one across from him. Bennett started to walk over, then suddenly stopped himself. "Actually, could I let Maurice stay here with you for a while so I can take a look around? He doesn't really like large crowds, and I had no idea it was going to be, well, this large. I just want to get a better sense of what you have going on here. Is that okay?"

Jim nodded, "Of course, happy to! He can have that whole sofa to himself, and I promise not to spoil him with treats. Well, not too many treats. He can have bacon, right?"

There was no point in asking because he'd already tossed several strips toward him, and Maurice caught every one of them, happily wagging his tail.

"Just don't be gone too long," Jim said. "You don't wanna miss the band."

"Wouldn't dream of it." Bennett scratched Maurice behind the ears. "Maurice, you be a good boy and stay with Uncle Jim. I'll be back soon."

"Have fun!" Jim called out as Bennett eased his way through the tent flap. Almost immediately, he was swept into the torrent of activity that was Market on Main.

• • •

It was all he could do to work his way past the concert crowd, but once he hit the stream of market goers, he found his pace. The tents were set up three across—one on each side of the street and another right in the middle—so that the crowd more or less followed the natural flow of traffic, heading south on one side of the middle tents and north on the other.

There didn't seem to be much organization to the vendors. One tent had crafts, while the next had artisanal vinegars; the next, jewelry; the next, a fine assortment of organic grass-fed beef. He saw several art booths, ranging from landscapes to abstract, and more pottery stands than he could count.

Something crashed into his foot.

For a moment, he thought Maurice had escaped from the tent and—miraculously—found Bennett in the crowd. But when he looked down, he was greeted instead by a robot.

A very *small* robot. It was no bigger than his fist and looked like a vintage toy car with funny robot arms.

"Sorry, Mister!" said a tiny voice, and suddenly a kid was running over to him with some kind of smart tablet clutched in his hands. "I'm still figuring out how this thing works!" The kid swooped the robot car off the ground and scurried off to a nearby booth before Bennett could respond.

Intrigued, Bennett surveyed the booth. Kids swarmed around two long tables, laughing and making things with their hands. Was it some kind of art project? Were they molding clay?

When he got a little closer, he realized it wasn't clay at all. It was tech.

The kids were controlling the robot cars. There were a couple of tablets lined up on the table like at an Apple store—though these were old and well used, not sleek and new and slate gray—and several kids were hard at work, typing code. The surrounding toy robots on the ground and tables looked a lot like the vintage car that had crashed into Bennett's foot. He saw a miniature Rolls-Royce Phantom, a Ford

Model T, and an Aston Martin Vantage, all with robot arms. Judging by the delighted squeals from the kids, it was their code making the arms move.

Bennett flashed back to the conversation he'd had with his uncle the day before. Hadn't Jim said something about a tech booth? He tried to recall his exact words, something about getting approached by a Silicon Valley type about a "mini Maker Camp for kiddos." Was this what Uncle Jim had meant? Bennett felt bad about how dismissive he'd been at the time. Jim had said "cool sound machines and lasers," but this was so much more than that. Seeing these kids' faces light up reminded Bennett how he'd felt when he first learned to code all those years ago: like he'd just discovered a whole new world.

His gaze was drawn to the other long table, where there was an entirely different kind of robotic character. This robot would absolutely be voted "most popular" in his high school class; more than a dozen kids were pressed together in the booth, trying to touch and talk to him. Upon closer observation, Bennett realized the platform interpreted sound, touch, light, and infrared inputs. From what he could gather, the outputs were LEDs and a speaker mouth.

Remarkable, he thought. Thanks to the eyes and mouth, the robot became an emotive type of character, more man than machine. Of course the kids loved it. It was actually talking back to them and responding in real time to their words and touch. Honestly, Bennett was twice their age, and *he* was mesmerized by the little guy.

Then he saw something even more mesmerizing.

A woman about his age was demonstrating how the tech worked. Bennett knew instinctively that the booth was hers; she moved easily between the groups of kids, leaning down to show them this or that or to type in a seamless stream of code. She was beautiful in a casual, carefree way, her auburn hair drawn up in a messy bun, her pale skin dotted with freckles and no makeup, her jeans and T-shirt casual chic. But even more compelling was the spark he saw in her eyes every time she answered a kid's question or typed a line of code. It was plain to see how much she loved this stuff and also how good she was at it. He felt like he could see her sharp mind at work.

Bennett started walking toward the booth. He got maybe 90 percent of the way there, then stopped short. His whole body went numb. Part of him wanted to strike up a conversation—and another part wanted to turn and run the other way. When was the last time he'd actually struck up a conversation with a woman? Did he even know *how* anymore? Add that to the list of things he was not: not able to charm a pretty woman, not even one who was possibly even more of a tech geek than he was—*especially* not a woman like that. Better to dissolve into the teeming crowd and never have to face her. It would be so easy. Just a few steps, and the current would carry him away.

But then he remembered something from the last podcast episode. Clear as day, he heard Mike Shapiro's words in his head. *If you're afraid to talk to someone you admire ... someone who you just haven't had the courage to strike up a conversation with, go do it! Go talk to a random person at a bar, or say hello to someone in line at the coffee shop.*

Or at the mini Maker Camp booth at your uncle's street fair, Bennett thought to himself.

Be open to the conversation, and you may be surprised where it takes you.

He took a breath. It was now or never. If he wanted to do things differently, to stop being a ninety percenter and at least try and touch the damn wall, then he had to take the first step.

The woman had just stepped away from the kids and was taking a sip from an ergonomic, blue water bottle when Bennett approached.

"I love your robots," he blurted, and instantly regretted it. What kind of weird pickup line was that? Your *robots*?

But her smile was warm and genuine. "They're lit, right? I'm so pumped the kids seem to love them. You just never know what kids will be into." She twisted the top back onto her water bottle and dropped it into her bag. "Maybe they'll wanna play with robots, or maybe they'll just want to keep scrolling on their iPhones."

"Nah, look at them!" Bennett gestured at the kids. "I can see the light bulbs going off in their heads. You're basically creating the next gen of coders who are gonna *invent* iPhones."

She beamed. "Or the next thing that's even better. Giving young people a chance to experience technology with hands-on demos is such an epic way to spark innovation. I love the moment when a kid sees the vast potential of what tech can do, you know? The digital landscape is just so *cool*."

Bennett found himself nodding along enthusiastically.

"Especially for girls," she continued. "There are still way more boys who go into STEM careers, but plenty of girls would totally dig science, technology, engineering, and math, if we just gave them the chance. I want to give kids that "Aha!" moment. It's kind of wild, right? Kids today are immersed in so much tech, but most of them aren't fully aware of how things work. I feel like if I can help get them to the point where they can understand how they can take a small piece of code and showcase it on a robot, that's a powerful experience."

She laughed, and Bennett thought he caught her blushing.

"Sorry," she said. "You didn't ask for me to download my whole 'Miss America STEM speech' on you. I just get so hyped about this stuff."

"No, I loved it," he said, because he genuinely did. "Honestly, I'm jealous."

She raised an eyebrow. "Of my robots?"

"Of how much you love what you do. I haven't …"—was he really saying this out loud to a woman he'd just met?—"I haven't felt that kind of excitement about my work in a long time. And the thing is, I used to get just as excited about code as these kids."

"You're a coder?"

He nodded. "App developer, yeah."

"Then you get it. Where other people see ones and zeroes, we see whole new worlds."

Bennett blinked. Hadn't he had that same thought, in pretty much those exact words, not five minutes ago?

"And to be fair," said his new mind-reading friend, "this isn't quite my *job*. Yeah, I used to code. I've always been good at it. I spent years working in the South Bay right out of college. But it kind of ground all the love out of me, and I came back home to Virginia last year to try to remember who I was. I started doing these mini Maker Camps just for fun. Because why should the Valley have a monopoly on innovation, you know? We've got so many brilliant, talented people right here, including a ton of kids who have the capacity. They just haven't been given the opportunity. If Google can put on a Geek Street Fair, I figure I can too. Anyway, I'm not jonesing to be a coder again, even though there were certainly things I loved. I'm glad I did it, and I'm equally glad I've moved on to the things I love now."

She laughed. "Look at me, spilling my whole life story. That was the long way around the block, but basically, yeah. My personal brand is a little different. I mean, don't get me wrong: it's still robots. If you want robots, I'm your girl. But I'm actually an artist. I've always been interested in the places where old tech and new tech collide, so my art explores that collision."

"Like making robots out of vintage toy cars," Bennett said, intrigued. "What kind of art? Do you have a portfolio I could see?"

When she handed him her phone, he felt a little thrill. The gesture seemed so easy and intimate, as if they'd known each other for years.

"That's some of my most recent stuff," she said, as he started scrolling through her online shop.

She wasn't kidding; if Bennett wanted robots, she really was his girl. Everything in her shop was some kind of robot. But unlike the more high-tech robots in the booth, these were mostly analog with a vintage feel, made from everyday items like spark plugs, lanterns, and gas cans. Sometimes they were shaped into functional things: coffee mugs and dinner plates, pencil holders and key chains, and— Bennett's favorite—a particularly cool laptop case made with gears and sprockets. But there were also straight-up art prints and sculptures. All the pieces boasted the same distinctive style, fun and funky. He'd never really seen anything like them.

"It's fun to make things people can actually use," she said over his shoulder. "Though sometimes it's also fun just to make art for art's sake."

"Do people really buy these?" He swallowed hard. "No offense."

She grinned. "None taken. You wouldn't believe how much I'm able to sell these for. Zoom in."

Bennett raised an eyebrow, then pinched the iPhone until he saw the price of one robot print. He almost dropped the phone.

"Two hundred dollars! People pay that?"

"Yep, they do. Quite a few of them, actually. It's all about knowing your audience."

She explained how most of her sales came from advertising on hyperspecific threads within news and open-format discussion sites, podcasts that focused on tech and robots, and even select viral videos on social media platforms. "There are actually more robot nerds around than you'd think. Though you do have to put in the time to source them."

"Sounds like you spend a lot of time on market research."

She sighed. "That's the thing. These days I spend more time online trying to find the folks who may be interested in robots than I do actually making robots! You get to the point where you wonder, *Am I an artist or a product marketer?* And I don't want to be a product marketer. I just want to make art—and to do these Maker Camps for kids, because I love it and it's fun. But I'm also someone who believes that tech can help us automate the parts of owning a business that are a major time suck. I spent years as a coder; I know what a good app can do. There's just gotta be some way to solve for *x*, you know—to figure out *where* robots are all the rage and then direct my marketing efforts *there*. But to tell you the truth, I'm only now at a place where I'm starting to think laterally. Feels like I'm finally finding my footing again. I'm actually going to the Ex Machina Expo tech conference next week in Charlotte to see if it sparks any big ideas."

She eyed him up, almost as if she was seeing him for the first time. "You wanna go with me?"

"Yes," he said, shocked to hear the word coming out of his own mouth. There were a million reasons to say no … and yet he'd said yes.

"Great. Meet me in Charlotte." She laughed. "Wow, I guess I just asked you out on a weird tech date, didn't I? You better not turn out to be an axe murderer or something."

He held up her phone, pointing to a robot who looked an awful lot like he'd been assembled of axe parts.

"I'm not sure I'm the greatest risk here," he joked.

She smirked and snatched the phone away.

"My name's Scottie, by the way. Not short for anything. I'm a girl, and my name is Scottie. End of story."

Something told him the story was only just beginning.

"I'm Bennett." He smiled and shook her hand, just as one of the kids started tugging at her elbow. She gave Bennett a little goodbye wave, and he knew it was time to be on his way.

In a kind of daze, he walked back to find his uncle. Had that really just happened? Had he just agreed to meet a total stranger at a tech conference in a few days?

And there was something else gnawing at the edges of his consciousness, something she'd said at the very end, about marketing her art. There was an idea stirring, though he couldn't quite see it yet. He felt like he was looking through a lens at a distant landscape that hadn't yet come into focus.

Inside the canvas tent, Maurice was very happy to see him. As he stooped over to scratch his loyal dog behind the ears, Mike's words echoed through his mind.

Be open to the conversation, and you may be surprised where it takes you.

To Charlotte, Bennett thought with a grin.

• • •

Even though the name of this podcast is *Read the Tape*, I haven't really told you what that means yet. I was getting you all warmed up. Now that you've been thinking about the last few episodes and applying some of the practices we talked about, it's time to talk about a skill that is central to succeeding in whatever it is you choose to pursue.

By the way, reading the tape isn't about literally reading—though you should be doing that too. Read everything you can. Read books, read articles. Read the *Wall Street Journal* every day if you want to be able to have things to talk about with people. Knowledge comes from exposure, and exposure is everything. I'm always exposing myself. I read everything I can, watch everything I can, take every meeting. Everything you could possibly do when you walk out of your house is another way of being exposed and gaining knowledge. That's what knowledge is: exposure! So read, read, read. But when I say "read the tape"? I'm talking about something different.

Bennett was intrigued. He switched on the cruise control of his truck and got comfortable in the driver's seat. Several days after Market on Main, he was an hour into his three-hour road trip to Charlotte. Back at his grandmother's house, Maurice was probably moping around, missing him. But Bennett would make it up to him with lots of walks and ball throwing once he got back.

He still couldn't believe he'd said yes to Scottie, that he was actually driving to a different state for a tech conference. He'd never been to a tech conference before. They weren't cheap—tickets for the Ex Machina Expo started at $2,000—and none of his prior employers had ever been willing to foot the bill. Frankly, he couldn't have afforded it this time, either, but Uncle Jim had made him a loan.

"No need to pay me back," he'd said. "Consider it an early investment in your genius."

But Bennett had every intention of paying his uncle back as soon as he possibly could.

He was a little nervous about seeing Scottie, sure, but he was mostly excited. Their conversation had brought something to life in him he hadn't felt in a long time, this feeling that he might actually be capable of creating something great.

Wow, he thought. He really had been implementing Mikeisms into his life. He never would be here, cruising comfortably down the interstate, if not for the podcast.

Now, with his phone connected to the truck's Bluetooth speakers, he leaned back in his seat, ready to hear more.

To "read the tape" refers back to the old practice of using a telegraph to transmit stock price updates. The machine on the receiving end was called a "ticker" due to the constant *tick, tick, tick* it made as it continuously printed out stock updates on long, thin strips of paper. If you've ever heard of a "ticker-tape parade," that's where the phrase comes from. In sky-scraper cities like New York City, folks would tear up bins of ticker tape and toss them out the window as a parade went by, the short strips fluttering down like confetti.

> Reading the tape is about reading the room, reading the environment, and reading the momentum of what's going on around you.

Reading the tape, however, isn't just about under-standing the movements of the modern world by staying aware of stock market fluctuations. It's much, much bigger than that. It's about reading the room, reading the environment, and reading the momentum of what's going on around you. It's about being able to take in visual cues, from the layout of a room to the raise of an eyebrow. It's about paying attention—accurately assessing people, situations, and trends—and using that information to do the right things at the right times in order to move forward.

I'll give you an example. Some years ago I was honored to speak at a prestigious MBA program at a Los Angeles university. Now, driving there that day, I felt like I had a good idea of what I wanted to say. But as I stepped up to the podium, I knew I needed to stop. I just didn't have a connection with the crowd, and I needed that in order to make my presentation resonate.

Thankfully, what I know about myself is that I can usually make a connection pretty quickly. So for a moment or two, I just read the room.

And what I noticed was that the people in the audience weren't too comfortable either. It was a typical university classroom with tiered seating, and I could see all the restless foot tapping, listless slouching, aimless doodling, and sneak peeks at smartphones going on around me.

I also noticed that everyone in the class had a sign on their desks with their name and occupation. There were veterans, midlevel executives, contract workers—all people looking to do something better with their lives.

And with that, I knew how I would start.

"We're going to start things a little differently today," I told them. "I'm going to go around the room, and I'm going to ask each and every one of you to state what you are afraid of."

They chuckled, but not a single one of them hesitated to share.

It only took about ten minutes for everyone to speak, but by the end of it, I had the beginning of a relationship with each and every individual. And with that connection established, I was able to speak, connect, and coach.

I told them my story—about how I failed and rebuilt, failed again, and found new avenues of success. I didn't hold anything back. It was full disclosure, completely honest and entirely transparent.

And it was amazing.

Out of the hundred or so students in the classroom that day, seventy of them reached out to me afterward, and I'm still coaching some of them today.

That's one way to read the tape. That's relating to people in a way that makes them feel heard and, consequently, helps you get what you need, because now you have that connection. It's about noticing what's on their desk, what pictures they have on their wall, the way they dress, the way they react, the books on their bookshelves (or lack thereof). It's about taking all of that in and interacting with them based on that quick but effective understanding of what means the most to them. Now you're not just some random entrepreneur out for your own gain—you've demonstrated that you actually care, and consequently, those around you often want to care back.

Reading the tape isn't just about reading individuals, though. It's about reading everything. If you're at a restaurant, for instance, you would read the tape by reading the people, the overall environment. Is it busy, quiet? Is there a massive table that your waiter is also trying to manage? What are your waiter's behavioral patterns? Are they rushed? Anxious? Frustrated? Engaged? Even something as mundane as their personal appearance. Is their shirt wrinkled? Did they wash their hair? If they've lost interest in their appearance, they've almost certainly lost interest in their work. Whatever is going on with them, you're going to have a better interaction with that server because you took the time to observe their circumstances and acknowledge what they're going through.

There was a long pause, long enough that Bennett checked his phone to make sure the Bluetooth hadn't disconnected. But then there was a soft *click* and the hiss of open air. Mike jumped back on.

I know I'm throwing a lot of high-level concepts out there, so let's slow the narrative down a bit and break this into smaller bites.

Let's start with the first time you meet someone. It could be the person interviewing you for a job, or it could be a random person at the bar. Whoever it is and whatever the circumstances, the first step to any kind of relationship is finding a way to connect with them.

This does not work if you start the conversation with your spiel.

You know what I'm talking about: your "comfortable go-to," your personal elevator pitch—the self-narrative we all have that explains to others why we're important or different or special.

"I'm an expert at this" or "I work for this incredibly well-known company" or "I'm cool because I'm aware of or involved in this obscure brand," et cetera.

All that is is a wall that makes it very hard for people to get to know you—the real you. Now, I understand that a lot of people out there may not want to be *known* when they first meet someone. They prefer the false front to the real thing. But if that's how you feel, then I really want you to think about this question:

Why?

You don't need me to tell you this, but sometimes we forget that there are no perfect people in this world. Everyone has flaws. Everyone has screwed up, and everyone has failed. You are not perfect, and the person you're meeting isn't either. So why try to introduce yourself as having it all together? Why not break down the wall and walk up to someone as you, not who you're pretending to be?

Connecting—*really* connecting—with someone means letting go of that false-perfect narrative and, instead of feeling like you have to prove yourself, being excited about getting to know them.

Open your ears and eyes, and listen in a three-dimensional way. See them as incredibly and imperfectly human as you. Notice their expressions, their reactions. Try to get a sense of their mood, their natural presence. Notice how they're dressed—if they're carrying a workbag, wearing sneakers with a suit, or have a child's handmade bracelet on their wrist. By focusing on them and not yourself, what can you learn about this person that will help you ask the right questions and connect with them as fellow human beings?

That's step one. Just be open and real, and let the other person lead.

Bennett reflected back on meeting Scottie a few days before. He hadn't given her his spiel or elevator pitch—maybe because, these days, he didn't really have one. But as a result, he'd inadvertently been more honest with her than he might have been before. And they *had* connected, on a real level, even surrounded by a bunch of noisy kids and robots at a street fair.

Still, it didn't take long for the fears to seep in. What if, over the course of the next few days, he and Scottie couldn't strike up that easy rapport again? They'd talked one time on the phone since the day they met, but only briefly, to talk logistics about when they were getting to Charlotte and where they were booking their respective Airbnbs. A part of Bennett had been dying to say, "Hey, why don't we drive down together?" But he understood that was maybe going a step too far, and he wanted to respect her boundaries. Besides, they'd have plenty of time to connect at the conference.

Once again, the podcast was eerily insightful, giving Bennett the exact wisdom he needed at precisely the right time. He kept listening.

Step two is *relating*.

As you take in all these observations, what can you ask that could help you relate with them? What sort of nonverbal communication could help you ask about their day? Do they seem down or excited? Are they anxiously drumming their fingers or standing with arms crossed? What could you say about your own recent frustrations, anxiety-inducing events, or joyful moments that would help you connect with them? You could also make educated guesses based on their attire. That handmade bracelet: do they have kids? The workbag: are they heading into an office nearby? Sneakers with a suit: maybe they're fitness buffs or regular runners. It could be something they said about where they're from, what they do, why they're in the same place as you at the same time. Inevitably, there will be something they say or something you notice that will lead to that deeper connection. All you have to do is stay open, listen, and ask.

Step three. *Ask the right questions.*

A lot of us are timid about asking questions. We're worried we'll hurt someone's feelings or make them feel awkward, but if you've done the other aspects of steps one and two correctly, then you already have a sense of how they like to communicate.

They may like to tell jokes or have a self-deprecating sense of humor. They could be an adjective junkie or like to talk in extremes. ("It was literally the *worst* day ever.")

Make note of whatever that style is, and adjust your response to complement it.

By the way, I don't mean mimic them. That would offend someone. Instead, if they're talking in a way that says "appreciate me and who I am," then ask questions with a complimentary vibe. ("That's an amazing return! How did you get into the day-trading world?") If they prefer humor, smile a little more. If they're sullen, adjust to supportive concern. Connect. Relate. Shape your questioning style just enough so that it complements the speaker's own style.

It's empathy but with an extra level of observation mixed with some educated guessing, and I'll tell you a quick story about that.

When I was in high school, I took the opportunity to earn some college credit by completing a certain number of hours of community service. So what did I choose? The local suicide hotline.

I know, not the most uplifting volunteer role. But it was and still is a real need, and it gave me an opportunity to connect and give back to people with a great amount of empathy, which is something I've always loved to do.

I drew upon my ability to connect with others on a deep, sincere level every second I was on that phone. I had never listened to someone so wholly or so intently before, but that's exactly what I found myself doing. Callers would tell me that they were hiding in the back of their closet, and I would just keep following these barely perceptible fluctuations in their moods, trying to find anything we could connect on that would help draw them back from that edge.

They needed to feel heard, so that's what I did; I made sure they knew I was listening.

And that's really what most of us are looking for in a conversation. We just want to be heard—not "heard" as in "repeat back to me what I just said" but "heard" as in "I get what you mean; I understand where you're coming from, and we see the same thing." The best relationships are built on that kind of authentic communication.

Now that we're starting to share some of that same headspace, we move on to step four:

Drop the hammer.

They've given you their practiced speech, you've adapted to it enough to make them feel comfortable, and you're doing your best to see the world as they see it. You don't have to agree with them, of course—as the saying goes, "The mark of an educated mind is to be able to entertain a thought without accepting it."

But we need to get past that wall, and it's going to take one more significant effort on our part: we have to articulate that we understand them in a way that they will understand it.

That's harder than it sounds because miscommunication happens all the time, even more so in our rapid-fire-response world.

This was another challenge of that suicide hotline. I may understand what they're saying and why they're feeling that way, but could I convey it back to them in a way that they believed me? That I not only heard what they said but understood what they meant?

That's what makes this wall such a challenge, but it's also why breaking past it is so rewarding.

"Tell me something about yourself that you don't normally share with others."

"What are your biggest fears?"

You know what? They're probably going to tell you. And for the first time, you'll see them as who they are and not as the polished image they present to the world. The wall may not be gone, but you've made enough of a break in it that you're able to make a real connection, and that's why they'll remember you. Because you were able to clearly demonstrate that you heard them. That's why they'll think of you before they think of one of your competitors, and that's how you grow your network of friends, supporters, and just old-fashioned good people to know.

When you learn to read the tape, not only do you make better, more lasting connections with others but you also learn how to read the bigger picture. What changes in behaviors, habits, needs, desires are driving fluctuations in industry and society? How can past trends, when viewed holistically, help us make realistic assumptions about the future and even the next big thing—

The podcast cut out abruptly. Bennett's phone was ringing. He glanced at it long enough to see Scottie's name flash across the screen.

He felt a flash of panic. It wasn't too late. He could turn around right now, claim that he was sick or had a family emergency. She wouldn't press—after all, they barely knew each other—and he'd never have to see her again. The impulse to turn around and retreat was fiercely strong.

And then he remembered Mike describing what it meant to be a ninety percenter. Bennett had felt so seen—and not in a good way. He'd been a ninety percenter in personal relationships, too, not just his career. He always got so close to connecting to a woman he liked, only to push her away. This pattern had plagued his dating life for years. He'd always chalked it up to the fact that he was a coder, and no one had ever accused a coder of having great social skills.

But suddenly he understood something. Every time he met a woman he liked, wasn't he afraid deep down that, if she really got to know him, he would only disappoint her? If he cut things off with Scottie now, before they went any further, he wouldn't have the opportunity to fail.

For the first time in a long time, Bennett decided the risk of failure was worth more than not trying at all. And the risk of success? Well, that terrified him, too, if he was being honest. But it was a terror he was finally willing to face.

He picked up the call.

"Hey," she said. "I'm driving down, and I know you are, too, so I was thinking ..." She laughed. "I don't know. I'm doing this all out of order, but I thought maybe we could get to know each other a little. You know, from the road."

Bennett smiled. All the other times he'd gotten to know a woman, he'd been so focused on his own fears. He couldn't get out of his own head. But maybe it wasn't about him at all.

All of the Mikeisms he'd just absorbed felt real and alive inside him. Could he do it? Could he listen, relate, and truly connect with Scottie? Could he be open and real … and let her lead?

A few days ago, he would have said no.

Instead he said, "I'd like that. In fact, I think I know how to start us off."

"Go for it," she said. "Shoot."

"Tell me something about yourself you don't normally share with others," he said.

And she did.

ABOVE ALL, TRY SOMETHING

___

CHARLOTTE, NORTH CAROLINA

Bennett stepped out of his Lyft onto the hot cement street of downtown Charlotte. He'd taken a Lyft from his Airbnb instead of trying to deal with parking at the convention center. Glass and steel skyscrapers rose up in every direction, and he didn't know which way to go. He felt suddenly like a country bumpkin.

"How do I find the—" he started to ask the driver, but the car had already pulled away.

The convention center, it turned out, wasn't hard to find. Everywhere he looked, people were walking toward it, some of them clutching briefcases or laptops or bike messenger bags, many of them holding Starbucks or Red Bulls or CBD iced tea—and all of them wearing the telltale lanyards and name badges of the "EX MACHINA EXPO."

So many people—young, beautiful, excited to be here, and completely at ease, like they went to these conventions all the time.

Those name badges—Bennett remembered with a sudden jolt the entrance ticket to the expo—were each worth more than $2,000. Every person wearing one was an app developer, a software engineer, or a product designer, just like him, someone who hoped to turn their idea into the next big thing and their $2,000 registration fee into a million bucks.

This is a lottery, he thought, *and every one of these people thinks they have the winning ticket.*

Bennett saw a queue of people standing by the cascading fountains outside the convention center and approached the end of the line. A cloud of cool mist rose up from the fountain, and he breathed it in, a nice reprieve from the afternoon heat. "Is this the line for registration?" he asked no one in particular.

A woman with black-frame glasses looked up from her phone and nodded.

"First time?" she asked.

"Yeah," he said. "To Ex Machina and to Charlotte. You?"

The woman shrugged. "This is my fifth time. I sold a start-up at Ex Machina a couple years ago, got another company into an incubator last year, and now I'm looking for an angel investor for my new company."

Her new company? How many companies did she have? She didn't look older than twenty.

"Anyway, good luck," she said, and then vanished into the crowd.

The registration line led to a set of kiosks, and next to each kiosk was a chipper kid in a bright-orange T-shirt. Bennett paused for a moment to read the tape. All the Ex Machina employees he'd seen so far wore these orange shirts, and this kid was college-aged, with a sprinkling of pimples to prove it. His face was full of eager shine and promise.

An intern, Bennett realized.

"Hi and welcome to Ex Machina! Are you here to register?"

"I'm preregistered," Bennett said, and pulled out his phone, looking for the confirmation email.

"Great!" said the intern. "If you have the Ex Machina app, just tap."

Bennett waved his phone at the kiosk and saw his name appear on its screen. It still amazed him, sometimes, how easy and almost magical technology was. He thought about everything that had to happen between his smartphone and that kiosk to make that simple "tap" work—how the two devices had to find each other on the network, do an encrypted handshake, pass packets of data back and forth, find Bennett's ID, and look up his registration in a remote database stored on a computer that was hundreds or maybe thousands of miles away—yet it all happened instantly and easily, with a "tap." This is why he loved technology: because it was like practicing magic. And somehow, the more he learned about how the magic worked, the more he loved it. It didn't make it less magical. It just made his magic more powerful.

"The printer's jammed," the orange-shirted intern announced. "We just called IT to come down and fix it."

"Can't you fix it?"

"Sorry, I'm just an intern."

Bennett felt a small flash of triumph that he'd read the tape correctly. "But isn't every single person in this conference in IT?"

"You'd be surprised! All kinds of people come to Ex Machina. I just met a lady who makes robot art."

Bennett's heartbeat kicked up a few notches. So Scottie was here. He realized that some part of him had been certain, absolutely certain, that she wouldn't come—even though they'd spent the last hour having an honest, real, connected phone conversation on the road.

Which in some ways made him *more* nervous. Now it would really hurt if she stood him up.

He looked at the giant digital wall clock: 3:59 p.m.

"Is it going to be long? The keynote is starting in a couple minutes, and I don't want to miss it."

But the intern just shrugged. "IT does what IT does, when IT wants to do it."

This was no good. He had to get in there for the keynote. "You know, I'm pretty good with computers. Maybe I can take a look and get it unjammed?"

"Oh, no. No one else is allowed to touch the printers."

The clock turned over to 4:00, and inside the main conference hall, thundering techno music began to play. The intern looked enviously at the door. "Man, the people in there are so lucky! Imagine being in the same room as the greatest tech innovator alive today!"

Bennett *was* imagining it. In fact, he had paid for it—and now he was going to miss it because this kid didn't know how to fix a printer.

"Hey, I'm just going to run to the bathroom, okay?" He didn't wait for the kid's answer. He stepped out of the line and started walking. He didn't know where the bathrooms were, and he didn't care, because he wasn't going to the bathroom. Instead, as soon as the kid looked away, Bennett walked quickly to a cardboard box that had been pushed out of sight under a table—full of empty lanyards.

He slipped it around his neck. It didn't have a name badge in it, but he could get that sorted out later. As long as no one looked too close, he could come and go as he pleased.

He made his way into the convention hall and took a seat.

• • •

The keynote speaker was late.

As Bennett watched various harried team members dart around the front of the event space while staring daggers at each other, he found it was almost second nature to read the tape. All he had to do was watch and notice. Deploying his most recent arsenal of Mikeisms, he noticed the team's expressions, reactions, and natural presence. He analyzed their body language and facial expressions, the way they were exchanging panicked whispers and gesturing backstage. He took note of their energy to get a sense of their mood.

As a result, Bennett understood what was happening long before the majority of people in the convention hall. The thundering techno music had been a red herring; the Ex Machina team was clearly trying to vamp while the speaker arrived from who knew where. Maybe his company headquarters. Maybe the moon.

Once it was clear to Bennett that he had some time to kill, he pulled out his phone and fit his AirPods into his ears. He turned up the volume, drowned out the techno, and tuned back into the now familiar voice.

> So you've made it to the next episode of *Read the Tape*. That means you must have liked the first few episodes well enough to come back for more. How's it going so far? Have you been reading the tape? What have you learned?
>
> I want to start today's episode with a brand-new Mikeism: *Everything comes back, but it comes back*

differently. Take any significant societal disruption, like a natural disaster or war. Afterward, when life begins to return to "normal," we notice that things aren't quite how they used to be. Some changes are obvious, and some are more nuanced. But when you get good at reading the tape, you start to ask the right questions about *how* things will come back differently. It's those questions that help us move forward, whether it's by making suggestions for improvement, coming up with new ahead-of-the-curve products in your day job, or launching a new business concept, either on your own or with a small start-up team.

> Everything comes back, but it comes back differently.

Reading the tape has gotten me into all kinds of interesting situations, some a little more precarious than others. But as I always say, you only have to be right 51 percent of the time to come out on top. That reasoning drives me to leap when others don't, to look where others won't, and to take risks when anyone else would have accepted the loss and moved on.

I'll give you an example.

For a few years, I ran a venture capital firm in Arizona. In 2007, a friend of mine approached me about opening a real estate office in California. He already had one in Arizona and said he was looking for a location with fewer seasonal fluctuations.

That, and he was looking for protection. In 2007, the Great Recession was looming, and the real estate industry as a whole was bracing for impact.

Still, I decided to look into it, and in the process, I found another real estate company called HOM in Newport Beach, California. One thing led to another, and I decided to see if we could merge the two companies. For a variety of reasons, we could not. But that didn't keep me from continuing to pursue investing in real estate, even though by this point the Great Recession was in full swing and real estate values were tanking across the country.

Why did I keep going after real estate? You're not the only one wondering. Just about everyone in my life was asking why in the world I was investing in real estate when the housing bubble had long since burst.

But, you see, I was reading the tape.

What's that Mikeism you just learned? Everything comes back, but it comes back differently. Real estate was going to come back, so buying in at the bottom of the market was great timing. As to *how* it was going to come back differently, a sharp eye reading the tape could see that while the recession was impacting Americans' decisions around home buying, international buyers were also seeing an opportunity.

That's right: I bought the majority of shares of HOM during the lowest point in real estate, which is always the best time to buy anything. We were able to turn the company around using that other evergreen Mikeism: pivoting. I read the tape and knew that it was the international buyers that were going to dominate the market. So we made a deal to become affiliated with a global brand and tapped into that international interest on the ground floor.

We became an affiliate of a global brand—as opposed to being a boutique local firm. That decision changed the course of HOM. At a time when most real estate was tanking, our company grew exponentially.

Now, something I haven't shared with you yet, even though my loyal listeners now know everything there is to know about my early years, is that I began my career as a market maker and trader in Chicago, ending up at the Chicago Board Options Exchange. Every day of that job started at one hundred miles an hour. We weren't just working in the markets; we were making them, creating liquidity for massive institutions that directly impacted the fluctuations of the market.

It was a phenomenally risky field with as much of a chance of catastrophic failure as there was for lucrative success. I thrived in that extreme envi-ronment, loving the rewards and hanging on by the skin of my teeth through the losses. Being at the center of the action and responsive to almost every conceivable news event was intoxicating, and as I learned how to follow the flow of events and stay several steps ahead of the next fluctuation, I was also getting better and better at reading the tape. Just as seeing the broader impact of a recession on international interest in US home buying was a way of reading the tape, so was watching the trend lines in hundreds and thousands of different sectors, watching to see what fluctuation would influence our margins and adjusting our spreads on a moment-to-moment basis, always staying ahead of the curve or risking ruin to the tune of millions.

After ten years, though, I was done. And not just because that kind of long-term intensity starts to age you in dog years. The role was actually evaporating as floor traders were replaced with high-speed algorithms and electronic trading formats, and my wife and I decided to move back to where we'd first met—Arizona—and reset our compasses, so to speak.

A lot happened in those years. I decided to go into the restaurant business, which I'll tell you more about in another episode. We also found out that my wife, Tara, had breast cancer. For a while there, we were traveling back and forth between Arizona and her treatment center in Orange County, California, living the dual coastal-desert life. When she was eventually—thankfully—given a bill of good health, we decided to go all in on the California coastal lifestyle.

Now here's where reading the tape took on its full, almost unbelievable potential—"unbelievable" if you didn't know what to read, but the literal road map to success if you do and you read it right.

I remembered when the dot-com bubble burst in 2001 and people were ditching stock in internet companies left and right. But that was short-term thinking. Yeah, a bunch of companies had overinvested in Y2K "preventative" software, and many of them ended up going bust, but a lot of them—more than most people remember—not only survived but grew at a phenomenal rate. That year, if you'd invested a little over $10,000 in Amazon stock, you'd be a millionaire by 2015.

So yes, while real estate was a tanking market in 2007, I knew from my marketing days that everything comes back. It just comes back differently. So I made the deals, got an office in Newport Beach, and began learning the industry while we waited for the sea change.

Now, there are pluses and minuses to learning an industry when it's at its worst. On the one hand, I had plenty of time and plenty of leeway to try different approaches, make mistakes, and learn from them. On the other hand, it wasn't cheap. At the bottom of it, the newly merged company was practically bankrupt with a measly $200 million in annual sales. But it was in the middle of this struggling market that

I noticed something about homeowners, at least, something about the homeowners in our particular area: they read real estate listings like they read the stock pages. The more properties around them sold for, the higher they valued their own net worth. Which got me thinking.

If property value was essentially a societal marker, then why not take a page from Goldman Sachs or Merrill Lynch and mimic their approach to posting stock fluctuations instead of trying to copy the marketing approach that other real estate companies used? They weren't exactly raking in the sales either. So we tried it.

Three years later, we'd grown from $200 million a year in sales to almost $3 billion.

Short story long, we ended up selling the company for a record amount, becoming something of a legend in the industry, with $7 billion in annual sales. And I'm proud of that because it never would have happened if I hadn't followed all those trend lines and read the tape.

As I'm recording this, there's another project we just started working on that relies on reading the tape in a whole new form. The idea is that it will read digital trend lines and mine data to give live, unbiased estimates on return on investment for home improvement projects. It's the perfect project for me because of my real estate background and my market-maker origins, as I already understand the relationships associated with equity markets.

When you think about it, housing is pretty much the biggest asset class in the United States. We have close to $50 trillion invested in it, and those individual values change every day. So it makes sense that homeowners would want to know what improvements, here and there, will give them the biggest bang for their buck when they do decide to sell, as well as *when* would be the best time to sell, based on all kinds of relevant factors and live data.

But I'm getting ahead of myself. Let's get back to the topic at hand: being willing to take a leap when others won't because, unlike them, you're reading the tape, watching those trend lines to see what will come back and how different it will be when it does.

There was something in what the podcaster was saying that made Bennett feel like something was just on the tip of his tongue. It was frustratingly close, like being on the edge of something thrilling, something world changing, but for the life of him he couldn't pull it out of his subconscious into a clear, cohesive thought.

Right now you may be thinking to yourself, *That's all great for you, but I have no idea what to look for, where to find it, or how to read the tape even when I do. How do I know if I'm even looking for the right thing, and how do I even know it's the right thing when I do find it?*

The thing is, everything is reading the tape. It's not like there's one clear trail of bread crumbs to follow; it's about training yourself to take in as much of the data around you as possible, from what someone says or does to the nuances of the environment and the flow of human interactions.

At the same time, it's not a step-by-step process. I can't give you instructions on how to read the tape because everyone observes things differently and absorbs data differently. What I can tell you is to be open to it. Everyone has their own innate ability to do this. It just takes learning how to be observant and to react to those observations in a way that stimulates positive change—whether that's seeing how a small but thoughtful gift can build instant loyalty with a teammate or how noticing an unmet need in the market could lead to a billion-dollar enterprise.

It's also about reading your own tape. What's your edge? What makes you tick? What about you is different from the rest of the world? What is your inner genius? Because there is something that makes you unique, whether you believe it or not.

And that's really what we've been trying to hammer out over the past few episodes, isn't it? What is it that gets you up in the morning? What comes easy to you? How can you observe the nuances of those things, big and small, that bring you joy and follow those tread lines to unearth your unique ability—your inner genius—eventually not only discovering something you love but something you could be successful doing? And don't give me that "There's nothing new under the sun" line. What did we just go over? Everything comes back, but it comes back differently. You just need to find that trend line that speaks to your passions and figure out where it's heading ... and get ahead of it. It's like what FDR said: "It is common sense to take a method and try it. If it fails, admit it frankly and try another. But above all, try something."

Now, in the past few episodes, I've challenged you to think about what brings you joy and to write a list that helps you see a little bit more into who you are and to do something that you're usually a little too scared to do, like introduce yourself to a stranger. Now, we're going to close out this episode with a challenge that builds on the last two and helps you to—what was that again?—right, read the tape.

Since we can read the tape on several different levels, let's do this exercise on two levels: personal and business.

Part one of the read-the-tape challenge: Find out something you never knew—and maybe something no one knew—about someone you admire. Bonus points if they say, "I don't think I've ever told anyone this!"

Part two: come up with a business idea that predicts the future.

I don't mean that you need to come up with a fortune-telling machine. I mean come up with an idea that will solve a problem that may not be on the table today, but it could be. It can be a local issue, regional, or even global. What trend lines are coalescing into a situation that you have the ability and passion to solve?

Here are some questions to ask yourself as you get started. What kind of tangible goods do we need to develop? What kind of transportation should we focus on as we move forward? How do we better manage supply-chain and distribution issues? What do future communication channels look like, and what do we need to get us there? What changes need to be made to the healthcare system? What other major trends and issues do we see or have we learned about that will impact the future?

Hey, I didn't say these challenges would be easy. But now that you're thinking about it, it's pretty fun, right? Now go, get out there. Learn something new, observe everything, look for those trends, unearth that inner genius, and keep reading the tape.

Just as the episode ended, the audience around Bennett erupted in applause—and the speaker strutted onto the stage.

• • •

An hour later, a whole convention's worth of software engineers and aspiring entrepreneurs poured out of the convention center's main auditorium and then scattered to all the various breakout rooms and panel discussions that were scheduled for the afternoon.

But Bennett walked in the opposite direction.

He and Scottie had agreed to meet up at a nearby hotel bar, on the logic that it would be next to impossible to find one another inside the enormous and crowded conference hall. Their plan was to go over the Ex Machina schedule together and strategize how they wanted to spend their time, to figure out how to get the most out of the conference—whether they should tag-team events so they would learn the same things and be able to discuss them together afterward, or whether to divide and conquer, cover twice as much ground, and then swap notes after the end of each long day.

He found himself nervously hoping they would decide to stick together, but he didn't know if that was because he thought it was the best way to learn from the conference or if it was just that he was really enjoying Scottie's company. Chatting with her during the drive to Charlotte had been so easy and natural, and they talked about every possible range of topics: favorite coding languages (she was partial to Python), favorite dog breeds ("All the big dogs, all of them" had been her answer), favorite ice cream flavors. Ice cream flavors was one of the few things they disagreed on. Bennett was a purist who liked the simplicity of a really good chocolate or a really good vanilla. Scottie swore by some abomination of a flavor she called "birthday cake," with sprinkles and frozen dough. "That's barely even ice cream," he objected, and finally they just had to agree to disagree.

It was so easy talking to her, in fact, that he had forgotten he'd only met her once, and as he walked into the bar, he panicked. Would he even recognize her? Would she recognize him?

But they spotted each other right away. She was perched on a barstool at an unusually high cocktail table. "Hey!" she waved, with— Bennett noted—a big smile.

He took the seat across from her. "Where's all your free swag?" he asked.

She rolled her eyes. "I'm not here to be a fangirl. We've only been here an hour, and now suddenly everyone at the conference wants to go to the moon, instead of whatever project they were working on when they showed up this morning. These conferences are exciting, but they can also be distracting."

She had a good point. It seemed like everyone in the bar was talking about space travel. That wasn't why they were here.

But why were they here? What project was Bennett working on? For a while now, something had been turning around in the back of his head, his own next big thing, but he hadn't quite been able to figure out what it was. He kept hearing Mike Shapiro's voice—and FDR's words—in his head. *Above all, try something.*

Bennett had come to Ex Machina for inspiration. He hoped being around all these people and all their ideas would help him understand his own idea.

"So, Mr. App Developer," Scottie said. "What are we doing next?"

That was the question. What was he doing next? How did he want to spend his time on this earth? What did he want to contribute? What problem would he be able to solve?

Before he could answer, a loud squeal of feedback filled the room. At one corner of the bar, a set of speakers were set up for—

"Karaoke?!" Bennett said, shocked. He couldn't believe people would come all this way to go to a tech conference and then waste time out of the middle of their day to sing karaoke.

Scottie looked surprised. "You want to do karaoke?"

No. No no no no no. Bennett did not want to do karaoke. He had never done karaoke. He had never in his life been tempted to do karaoke. He didn't even know how to do karaoke.

So he was as surprised as anyone when he answered "Yes!" and grabbed her hand and headed toward the stage.

• • •

Bennett, it turned out, was terrible at karaoke. But it also turned out it didn't matter to him one bit how terrible he was, because it was fun. Karaoke was fun! Who knew? He and Scottie belted out an atonal rendition of one Ben Folds song after another.

"I don't think I've ever confessed this out loud to anyone," Scottie said, as they took a break between songs to guzzle two beers, "but Ben Folds is probably my favorite musician of all time. He's so good I think he should be knighted."

Bennett grinned. "Sir Folds Five."

"Laugh if you want. But I'm telling you: the guy is a megastar."

Only as Scottie grabbed Bennett's hand and dragged him back onto the stage did he realize what she'd just said was a variation on *I don't think I've ever told anyone this.*

He couldn't believe it. Had that really happened? He had done Mike's first homework exercise, simply by saying yes to karaoke.

Scottie's cheeks were flushed from the beer, and Bennett felt a little tipsy himself as he reached for the mic. Could he hit the high notes? No, he could not. But people in the bar were singing along, too, helping him out, and when they laughed, they were laughing *with* him instead of *at* him, and he realized this was the best he had felt in a long, long time.

How did I get here? he wondered as he gazed around the room—a tech conference in Charlotte, next to a new friend and maybe business partner and maybe something else, on the cusp of new opportunities and new adventures. *How did I get to feel this alive?*

What was it his Uncle Jim had told him? "When you really decide what you want to do and you put everything you have into making it happen, chances are you will. And if you don't, hey, it's better than never trying at all." That was only a few days ago, but it seemed like another lifetime.

Another remarkable thing had happened while he was onstage singing: his idea had happened.

It happened right at the moment that he had stopped thinking about it—stopped thinking about needing a new app idea, stopping thinking about coding or entrepreneurship or what he was doing with his life or, more accurately, what he was *not* doing with his life. It had happened when he stopped thinking altogether. He was singing the chorus of "Brick," that high note that he couldn't hit. He closed his eyes and sang and felt the music carry him away a little, and out of nowhere, like a gift from the gods, it popped into his head.

And just like that, the idea that had been gnawing at Bennett for days sprung loose. He knew what he was going to do.

He knew how to predict the future.

"I want to talk to you," he said to Scottie as soon as the karaoke song was over, "about a new app."

He didn't realize his mic was still on until a drunk heckler shouted out, "Tell us about your app!"

· · ·

They found a quiet corner of the convention hall where they could talk. They were missing the afternoon's panel discussions, but Bennett didn't care. Right now, the only thing he cared about was his idea for a new app.

"You said you spend more time trying to market your robots than you spend building robots, right?" he asked Scottie.

She nodded vigorously. "Yeah. It drives me crazy how much time it takes!"

"What's the time-consuming part?"

"All of it! I have to find places on the internet where people are into robots, or art, or especially robot art. I do a bunch of searches and find websites or forums or Facebook groups. Then I have to see if there's a way to buy ads in those places, to reach those people. But every site has a different ad platform—so if I want to reach the Facebook people, I buy ads there; and if I want to reach Twitter people, then I buy there; and if I want to reach that random flea market I found in Saugus, California, or wherever, then I have to see what ad platform they use, if they even use one. Everything is different, nothing is standardized, and I have no idea what's working and what's not working. It's too much."

"It's too much," Bennett agreed, "because it's too big. The internet is too big. There's no way you can possibly find all the places you want to reach."

"Exactly! And even if I find some place to advertise, who knows how many I'm missing? What I need to know is, Who is talking about robots, or robot art, right now? How many websites ... how many *communities* are out there who would be really into what I'm doing, and I'll never even find them?"

"So what if you could find them? What if you could find *all* of them?"

Scottie laughed. "Well, then they still might not buy my art, because not that many people buy robot art. But," she conceded, "if I knew I was reaching them, if I knew I was at least getting in front of all the right people, then I'd have a chance, instead of just, you know, putting messages into bottles and lobbing them into the ocean—which is what it feels like right now."

"So let's find them," Bennett said.

"Believe me, I've tried. It just takes so much time!"

"When we met, you said you should be able to 'solve for x.' Well, this is x. This is what we need to solve for. This is what we need an app to do. We're going to create an app that can find the robot-art people—find them wherever they are, in whatever corner of the internet, and leave no stone unturned. With machine learning, it will get better and better at finding exactly what we need—and when it finds it, it can scrape that page to learn what ads are on the page, how the ads get served and with what parameters. It can do all this monotonous, time-consuming work, while you get to focus on mini Maker Camps and making art."

Scottie was silent, no doubt thinking about how such an app could work—what problems they would need to solve, what problems they would discover that she couldn't yet imagine.

"You have questions," Bennett said.

"So many questions," she agreed. "But only one that really matters."

Bennett felt a pang of butterflies, suddenly afraid she had found some crucial flaw that would put a hole right through this idea and all his enthusiasm for it. He swallowed hard and then said, "What's your question?"

"Can we also teach the app to make my morning coffee?"

• • •

By that evening, they had cocktail-napkined their idea into something that might actually work.

Writing a script that could crawl every corner of the internet was easy enough; bots and spiders had been around as long as the World Wide Web. But the parts of this app that were hard were really hard.

First, they wanted to create a bot that could seek out the content it was interested in—in this case, robot art—and ignore everything else. Later, in future versions of the app, they would be able to teach the bot to search for anything they wanted. In fact, Bennett was already imagining that once he built this prototype, he would have a piece of software that he could repackage and resell over and over, for any industry. But for this version, for Scottie, the app would look for robots.

The tricky part: they wanted this bot to get better and better at making its own judgments and deciding for itself what was and wasn't robot art. Bennett wasn't even sure that *he* could give a clear answer about what makes something "art" or not—and he definitely wasn't sure he could teach a computer to decide. But computers were great at learning patterns, and he would program it to look for things that fit the pattern of robot art.

The second big puzzle they would have to solve is, once their bot found robot art, then what? Scottie had described her frustration: so many web pages have ads—but figuring out how to get your own ad on any particular page was a total crapshoot. Her main method so far had been to look for a "Contact Us" link on each website and then write an email by hand asking if she could put her own ad on the site.

"That's worked about as well as you might expect," she said.

"How many people say yes?" Bennett asked.

"No, what you really want to know is how many people even respond to my message." She held up her hand and put her index finger to her thumb. "Zero point zero. Zilch. Nada."

Bennett opened up a few of the websites Scottie had been trying to reach and looked at their HTML code. Wherever the site had an ad, the code was a littered mess.

"What are you doing?" Scottie asked him.

"Each of these ads comes from an ad server, right?" That's how it usually worked: the web page they were looking at had some embedded code that would send a message to some other faraway computer, saying, "I want an ad here!" Then that faraway computer—the ad server—would send an ad to be displayed on the page. What kind of ad—what size and shape and what content—was all embedded in the metadata that passed between the two computers.

Scottie saw what he was thinking. "You're thinking we can scrape that metadata?"

"Yeah," Bennett said. "Our app scrapes the metadata, learns all about the ads on this page—what they are, their size and shape but also their content, the demographic they're trying to reach, that sort of thing. We can learn everything about what sort of people visit a site by scraping its ad metadata."

He could see wheels spinning in her brain. "But there must be dozens of different ad servers, hundreds. And there's no standardization. Every single ad server is going to use code that's a little different than the others. We're going to have to teach our app to understand all of it."

"Then we'd better get to work."

DON'T LIMIT YOURSELF

SOUTHWESTERN VIRGINIA

A week had gone by—an exciting, exhilarating, exhausting week.

During his drive to Virginia from Charlotte, Bennett realized with a laugh how little of the Ex Machina conference he had actually seen. Coders kept coming and going from the hotel with ever-changing bags of new swag—pens and T-shirts and toys and gadgets—but Bennett mostly stayed in the hotel bar, drinking coffee and brainstorming with Scottie about their new app.

Bennett didn't even feel bad about missing the conference. The reason he had gone, after all, was to find inspiration—and boy did he find it. Once he started thinking about the app, he couldn't stop: his brain kept going over the various puzzles and how he would solve them, and he would explain his solution to Scottie with words that tumbled out of his mouth, only half making sense, and she would ask questions or point out things that would set him on better solutions or more puzzles that needed to be solved.

He didn't want to waste a single minute. That's why, a week later, he looked so rough. He was unshaven, in rumpled clothes, eyes strained from so many long hours staring at his computer screen. He would have worked around the clock nonstop if it weren't for Maurice showing up every few hours demanding to go outside. And even while he was walking the dog, Bennett was thinking about code. Luckily, they were being well fed by his grandmother, who plied them with her famous pancakes and all sorts of fine southern fare. Gran had taken a real shine to Scottie, who seemed to like her just as much.

Bennett had barely noticed. As of today, he had a working prototype of the app.

He was calling it "RobotFindr." He tried spelling it the normal way, *RobotFinder*, but he was sure that it looked more app-like when he got rid of that *e*.

"Like Grindr," Scottie said. "Matchmaking, but for robots."

"No, it's nothing like Grindr!"

"Relax, Bennett. I'm just joking." Scottie had an ease about her that made her a great work partner. Before they had even left Charlotte, they had decided this would be Bennett's project. Scottie claimed her programming skills were too rusty for her to be much help, but Bennett had a suspicion she was just being modest. She had been working in Silicon Valley, so, no doubt, she was the real deal.

The truth of it was, she just didn't want to write code. "We're doing this so I can spend more time making art, remember? And so that you can write the next big thing." Bennett agreed.

Scottie had been fantastic about making suggestions to improve the app, and never in a way that made Bennett feel bad about his own idea. If he was getting short-tempered with her, then it was a real sign he hadn't been getting enough sleep.

"Sorry," he mumbled. "It's been a long week."

"No big," she shrugged. Then she held out her hand expectantly. "So … can I see it?"

The question caught him a little off guard. He knew he was going to have to show the app to people eventually, and he was also really proud of what he had done. But the idea of letting her see it made him suddenly self-conscious. *What if it's terrible? What if I've been doing it all wrong?*

He heard Mike Shapiro's voice in his head: "Above all, try something." He gulped and handed Scottie his phone.

Much like Bennett, the app looked rough around the edges. There hadn't been time to make it look pretty. The app's launch screen was just a big red button that said "Go!"

"It's not much to look at," he warned. "The real magic happens once you push the button. That tells the app to start crawling the internet, hopping from link to link to find relevant content."

"Robots?" she asked.

"Robots," he agreed.

She pushed the button.

In a moment, the screen started filling up with text—arcane, unreadable text. Data.

"Wow," Scottie said. "You weren't kidding. It's not much to look at."

"It's just a prototype," he said. "I needed to be able to see—"

"Yeah, I know," she said calmly. "We need to be able to see what the app sees. Defensive much?"

Bennett sighed. "Did I mention it's been a long week?"

She smiled at him and looked back at the screen. "So whenever it finds a web page with content about robots, it scrapes the page for metadata. That's what I'm seeing? The metadata?"

"Exactly. The code looks at any ads that appear on the page and tries to extract useful data from the ads. If I can teach the app to read the metadata, then we know everything there is to know about the demographics of the people who visit this site."

She nodded. "And then I know exactly where to focus my own marketing efforts. This is really good, Bennett!"

He could tell she meant it. He didn't know how stressed he had been all week until he felt the stress lift off him, like he had just taken off a backpack full of stones.

"So what happens to all this data you're collecting?" she asked.

And just like that, the stress dropped right back into his shoulders. "I gotta hand it to you, Scottie. You have a real knack for asking the hard questions."

She laughed. "That's why I'm here."

"What to do with all this data is my next big problem. So far, all the data I've been collecting is getting dumped onto my laptop, and that's where I run the machine learning so it can get better at sorting the good stuff from the bad stuff."

"You're about to say, 'But …'"

"But … the job is way too big for my laptop. The only way RobotFindr works is if we have lots and lots of data. We need big servers."

"And big servers need big money."

Bennett nodded. "Or at least some money—which is more than I have."

He slumped down onto the sofa. He had known this was coming, that sooner or later he would need to scale the app and start laying out some cash. But he had been so busy writing the code, he had forgotten to deal with this impending problem. Forgotten? No, not really. More like, ignored. He stayed in his comfort zone, writing code, doing the thing that made him feel good, instead of facing the problem that he knew he would need to face, the thing that would make him feel bad.

A feeling washed over Bennett that was miserably familiar. He had fallen right back into the trap of being a ninety percenter. He was coasting, doing the comfortable thing, the stuff he was already good at. For the first time, it occurred to him that coasting was really a subtle form of self-sabotage. Because as long as he stayed in the world of code, he would never have to take the app to the next level—the funding level.

The truth was, he needed money, and he had none.

"So what's next?" Scottie asked. "Can you write an app called MoneyFindr?"

"Ha ha." He was demoralized, but he knew he couldn't let this setback get him down. "Okay, MoneyFindr is sort of funny."

> For the first time, it occurred to him that coasting was really a subtle form of self-sabotage.

"Crawls the internet, looking for people with money ..." She smiled impishly—and Bennett got an idea.

"Where are you going?" Scottie called as Bennett left the room.

"I'm looking for someone with money!"

• • •

Uncle Jim was in his driveway, washing his shiny, new Ford F-150.

"Uncle Jim," Bennett said in greeting. "That truck sure looks good in your driveway."

Jim looked up and waved. "Howdy, son. Good to see you. You know, Emma sure does love that new car of hers, but it can only do so much. This beauty here? Practical *and* comfortable." Jim turned off the faucet to his hose. "How did things go down in Charlotte?"

"Great! I can't thank you enough for sending me. In fact, that's what I came here to talk to you about."

Jim motioned to two empty Adirondack chairs, and they both got comfortable on the big wraparound porch. Bennett had to chuckle, remembering how intimidated he'd been by this house before knowing it belonged to his uncle. That was less than two weeks ago. Amazing how much could happen in such a short amount of time.

Bennett told Jim about Scottie, about getting the idea for his new app, and about how he had spent the past week building out the prototype. He opened up his laptop to show his uncle the sketches and the code. "It's good, Uncle Jim. And it's ready to go to the next level. All I need is a little help."

Uncle Jim was quiet. He reached for the pitcher of lemonade and refilled their glasses, saying nothing, and put the pitcher down. "I believe in you, Bennett. You're a good kid, a smart kid, and, judging by the week you've had, a hardworking kid too. It sounds like this app could have some real promise."

He picked up his glass of lemonade and drank a gulp of it, savoring it as it went down and pausing for what felt to Bennett like a year and a half. "So, does that mean you're in? You'll be my first investor?"

Uncle Jim put down the lemonade and looked Bennett in the eye. "Because I love you, because I want you to do big things, I'm going to say thanks, but no thanks."

Bennett blinked. He wasn't sure he'd heard right. "You're saying no?"

The older man shook his head and gave him a look that was hard but not unkind. "I'm saying that you and this app are promising enough that you should go try out for the big leagues. Stop acting like a kid, hitting your uncle up for money, and start acting like a businessman with a real product to sell."

Bennett nodded, because what his uncle was saying made sense, but inside he was screaming. Uncle Jim was saying no? "Uncle Jim, I would love to do that. I would love to act like a businessman with a product to sell. But I don't know how. I don't know how to approach a venture capitalist. I don't even know where to *find* a venture capitalist!"

Uncle Jim met his gaze but didn't say anything. Why was he doing this? What point was he trying to make?

"You are literally the only investor I know. You're probably the only investor in this half of Virginia. It's not like we're in Silicon Valley!"

Uncle Jim shrugged. "Eventually, the little bird gets pushed out of the nest, and it has to learn to fly."

"What does that mean? Fly? Fly where?"

Uncle Jim reached for Bennett's laptop and typed into a browser window. When the web page loaded, he turned the laptop back toward Bennett.

SILICON VALLEY PITCH FEST

Featuring special guest host Mike Shapiro!

"What if you fly *here*?" Jim said.

Bennett was shocked. "Mike Shapiro is going to be there?"

"One night only!" Jim laughed. "He's going to be a guest judge on that show *The Lions' Den*, and they're taping an episode this week in Silicon Valley."

"And you just happened to notice?" Bennett couldn't believe his good luck.

"No, kiddo." Jim winked. "I just happened to go to mikesshapiro.com."

• • •

A few days later, Bennett was on a plane.

It hadn't been easy. The entry fee for the Pitch Fest had almost maxed out his credit card, forcing him to buy a red-eye to San Francisco with what meager credit he had left. He understood what Uncle Jim was trying to do—he even begrudgingly respected him for the tough love—but as Bennett boarded a plane at 11:50 p.m. for the first of two three-hour flights, with a layover in Denver, he found himself feeling small and stupid that it had come to this.

The most embarrassing thing was that he had asked Scottie to come with him, then froze in terror when he realized she might think he was offering to pay her way.

In the end, it didn't matter. She had gently but firmly refused.

"I just *left* the South Bay, remember?" She shook her head. "No, Ben. This one's all you. Besides, you gotta get out there on your own to see what you're made of, right? Dazzle them with your app. Show them you're a genius. Or"—she'd given him a wry grin—"show them your *inner* genius." Thanks to Bennett's recommendation, Scottie had been listening to *Read the Tape* too.

But at 2:00 a.m., when he hadn't slept a wink and still couldn't get comfortable in his middle seat at the back of the plane, he didn't feel like an inner or outer genius. He felt like a fool.

What was he thinking? Taking RobotFindr to the next level would take far more resources than he had available to him or could ever afford. But he had pushed that aside, overtaken by the excitement of working on something he believed in and that could truly help others. At the time, it seemed exciting, even noble.

Now it just seemed futile.

He dropped his head into his hands.

"What am I going to do?" he mumbled to himself, earning a sharp glare from the lady in the aisle seat. There was no way he could pull this off—no way he could get the money he needed, the talent it would take to build this. As Scottie had pointed out, his app wasn't much to look at. If he was going to do this right, he'd need an app designer, and with no money and no investors, who was going to take a chance on some guy from Virginia with his robot app?

A particular Mikeism came to mind: *You only need to be right 51 percent of the time to be successful.* And if Bennett *didn't* try—well, then he was guaranteed to fail.

Frustrated, he pulled out his phone and scrolled around for something to listen to. His first instinct was to play soothing music to help him sleep. But when the *Read the Tape* icon flickered across the screen, he clicked on it. It had been a while, and right now, he could use some encouragement.

> Welcome back to *Read the Tape*. In this episode, we're going to talk about the importance of not limiting yourself.

Bennett almost laughed out loud—but caught himself just in time, lest he make Aisle Lady his sworn enemy. How was it Mike always knew *exactly* what he was thinking or worrying about? Was Bennett listening to the podcast, or was the podcast listening to him?

Up until now, we've talked about your attitudes and beliefs. We've discussed how changing your perceptions—about yourself, others, and the world around you—can open up possibilities and guide you toward discovering those things that drive you, make you unique, and help you achieve success. Throughout, those qualities that make you who you are—your skills, your experiences—have been integral to these discoveries.

Now, we need to talk about pushing past what we know about ourselves and dive into what can be.

Often and usually without realizing it, we let our experiences hold us back instead of propel us forward. We box ourselves in, defining ourselves as what we have done—those lines of "skills" and "previous roles" on our résumé—instead of what we could do.

> But what if, instead of thinking about what our experiences say we can do, we look at what we're capable of doing?

In doing this, we risk burying ourselves in the tedium and mediocrity of settling for work that we already know, which numbs our minds and souls even if the paycheck is decent. We don't push ourselves, and instead of growing and pursuing new avenues that could spark excitement and reignite our happiness, we instead become complacent, bored, and lost.

But what if, instead of thinking about what our experiences say we can do, we look at what we're *capable* of doing?

While experiences say "I'm able to do X, Y, Z skills," capabilities open up the door for anything we believe we can do. Maybe you've never gone snow skiing before, but you likely believe you're capable of doing it. You may even believe you'd find it enjoyable. Like confidence, believing we're capable of doing something simply means that although we may have never had the experience or developed the skills to make it happen, we believe we can still try it and learn how to do it—and we may even find that we can do it pretty well.

If you recall, I started a story in our last episode about owning a restaurant. I'm sure you've been very curious about what happened to that. It also happens to be a good example of what we're talking about today: reaching out beyond our experience to find out what we're capable of doing.

This goes back to before my real estate days, when I was still a trader and investing in businesses in

Arizona. But I was still looking for something else to do, something I could really invest in and help grow.

Around the same time, we were regulars at a restaurant not far from our home, which was run by one of the most celebrated chefs in Arizona. And because we were there so often, we ended up becoming friends with the chef and his wife. One day I made an offhand comment about how he should let me know if he ever wanted to do something else because I'd be interested in investing in it. Before I knew it, I was helping him open a restaurant.

At first, I'd told the chef that I'd fund a certain amount, and it was up to him to raise the rest. But after I heard some of the offers he was getting (and how much interest those investors would retain), I offered to fund the whole venture on my own.

His idea was to make it a high-end modern Mexican restaurant with a focus on Oaxacan cuisine. I loved it, but the idea turned out to be ahead of its time for the area. It also turned out that the chef—who I put in charge of the whole enterprise—simply wasn't capable of running the restaurant on his own. However, I wasn't going to find that out right away. Instead, shortly after opening day, my wife was diagnosed with cancer, and I no longer cared how the restaurant was doing. I only cared about her getting better.

When she finally, thankfully, recovered, we came back to a train wreck.

We knew it was going to be an expensive restaurant, but what we'd failed to take into account—where I failed to read the tape—was that Mexican cuisine in Arizona was considered to be inexpensive food at that time. Even if it was an exotic, well-prepared selection, it was too much of a disconnect to draw in much of a crowd, and it was losing money hand over fist.

One thing led to another, and the chef left, leaving me with two options: make it work, or risk losing millions on a failed venture. And I wasn't going to lose money on this.

Deep down, even though I hated the situation, I loved the challenge. I had to fix things—and I had to fix them now.

This time, I didn't try to force an expensive twist on an inexpensive cuisine on the public. Instead, I listened to what they wanted. I poured through the negative reviews in local papers to see where we were failing, I read the trend lines in the restaurant industry as a whole and how it was trending regionally and decided to refocus on a more casual Tex-Mex style with affordable prices. At the same time, I didn't want the place to blend in with every other Tex-Mex restaurant in the area, so I recruited a chef who specialized in sustainable dining, which was also an increasingly popular approach back then.

Between the lower prices, more recognizable dishes, and the appeal of sustainable dining, we were able to turn the restaurant around. And while it wasn't wildly profitable, it did well enough that I was able to sell it for a price that more than covered what I'd poured into it. In fact, the building was so valuable we were able to make a large profit.

What did I learn? That I was capable of running a restaurant. In fact, I enjoyed it, but it was a full-time gig, and I didn't have the kind of time necessary to run it well indefinitely. I learned what it was like to be a complete and utter failure at something and then what it was like to turn that failure around. It was hard to swallow at the time—those newspaper reviews were especially harsh—but I loved seeing what it became once I finally took a step back and read the tape.

Time and again, studies have found that part of human happiness depends on our willingness to be open to and learn new things as our lives progress. Pushing the boundaries of what we believe we're capable of renews our joie de vivre and teaches us to trust ourselves.

One place I do this is the gym. I've got a small gym in my office because fitness is so incredibly important to my daily life. My father died at fifty-five, when I was in my midtwenties. My father's father died at fifty. That's an element of why fitness is important to me; for me, exercise is a correlative behavior associated with living. I recently had a major health event myself. Everything's fine now, but it puts things in perspective. I think I'm somewhat obsessed with exercise for health-related reasons. I want to live.

One thing I teach is that you need to find your safe space. Where do you clear your head? Where are you most creative? For me, the gym is my space and where I clear my head. So yes, I do spend an inordinate amount of time working out. That's because of the health benefits, but it's also where I get creative and do some of my best thinking. If you've ever found me on social media, you'll see I record whole videos from the stationary bike.

Keep moving. Don't get stuck. Move all the time. Go outside. Exercise your brain; exercise your body. It is absolutely mandatory. If you're sitting on a sofa, what's going to happen? Nothing. Moving keeps your head clear, reduces stress, and makes you more successful in every endeavor. It keeps you moving *forward*, pushing yourself, pushing your boundaries. When you're lifting weights at the gym, it just might inspire you to take on more "weight" in the other aspects of your life.

Think of it this way. When we're young, the world is a new and amazing place. We soak up information, learning languages, musical instruments, anything we can read about or put our hands on. But as we get older, the stresses of life's responsibilities hold us back, and we resort instead to the comfort of the familiar instead of the risks of something new.

But when we venture to learn something new— whether it's going to the gym or learning to surf or cooking a new kind of cuisine—we find that our minds are stimulated and refreshed, our confidence is bolstered, and our comfort with risk is strengthened.

This is why success can't be found just in your experiences, in what you've done. Only when you look toward what you're capable of doing and what you're willing to learn can you find the real joy and happiness of accomplishment. If you are doing things you like, you will be successful. Success starts with doing something you love.

That hits pretty close to home, Bennett thought, thumbing the pause button on the podcast. He glanced around the plane. The cabin lights had been dimmed, and everyone seemed to be sleeping, including his seatmates on either side. He wondered about their life circumstances and what had brought them here, crammed into coach on this red-eye flight to Denver. Were they all ninety percenters— people who got most of the way there, so close to the success they wanted, but then didn't have what it took to touch the wall?

He felt a flash of resentment. It was all well and good that Mike Shapiro was off somewhere in a faraway recording studio, urging listeners not to limit themselves, to try new things. Mike who worked out in his office gym every day, while Bennett couldn't even throw a ball to his dog without getting winded. Or maybe Mike was already in San Francisco, hanging out with the other Lions and prepping for his special guest appearance on *The Lions' Den*. But through all his examples, no matter how hard the challenge, it always seemed like he had some source of funding to draw from. And Bennett did not have that luxury.

Or was he just limiting himself?

Because here was the thing: he was going to Silicon Valley Pitch Fest to *find* funding, to demo his app for potential investors and get people excited about what it could do. If it was true what Mike said about human happiness depending on our willingness to learn new things, to push our boundaries of what we're capable of, then Bennett was on the right path. He might even be on the right plane.

Only when you look toward what you're capable of doing and what you're willing to learn can you find real joy.

What was Bennett willing to learn? And what was he capable of doing?

He pressed his head into his seat back and shut his eyes. For now, he hoped he was capable of doing one thing: sleeping.

And he did.

KNOW YOUR NUT

SAN FRANCISCO, CALIFORNIA

The sun was coming up as Bennett's plane began its approach to SFO airport. He had never been to San Francisco—had never been to California at all—and he tried to peer through the window from his middle seat, gaping at everything he saw: the blue, choppy water of the San Francisco Bay, the flare of golden sunlight bouncing off the city's downtown high-rises, the iconic enormity of the Golden Gate Bridge. And, as the plane looped around for its landing, Bennett also saw the vast Pacific Ocean stretching all the way to the horizon.

California, here I come, he thought, and smiled.

This all felt so far from his normal life—in miles, yes, but also from the life he had been living just a few weeks ago. He was in California to pitch an app idea to his new hero, Mike Shapiro. It didn't seem real.

But it *was* real, and as soon as the plane touched down on the tarmac, the reality of it hit Bennett hard. In a few short hours, he would be presenting his software to Mike Shapiro and a panel of other judges for a special episode of a TV show called *The Lions' Den*. The show featured five successful investors listening to the pitches from different contestants. Usually, it was filmed in Southern California and included pitches from all sectors of business—but this week, *The Lions' Den* was doing a special episode on Silicon Valley.

The format of the show was strict: each contestant had exactly ninety seconds to describe their idea. Most of the people who went on the show made their pitch in one long, smooth, carefully rehearsed speech, and many of them were able to finish just at the moment that the ninety-second timer buzzed.

After the buzzer, the camera would cut to the judges—the "Lions"—and the fun would begin.

The Lions' Den was aptly named; the Lions weren't known for gentleness. Bennett had seen a few episodes of *The Lions' Den*, and they had been brutal affairs. Contestants were often reduced to tears. A project that someone had worked on for years could be torn to pieces in moments, dismissed with an eye roll or an offhand quip. A few of the judges seemed to genuinely relish finding cruel things to say and would sometimes try to outdo one another with meanness, spitting out one cutting line after another.

It could be terrible to watch, and for a moment, Bennett felt his heart race knowing that, before the day was out, he would be the one in that hot seat. What was he thinking? He didn't belong here! This was Silicon Valley, where all the world's best software developers came, where entrepreneurs pitched to investors every day, three or four times before breakfast. Bennett was confident that RobotFindr was a good idea, and he had a working prototype. But he wasn't at all confident he knew how to sell it.

Then again, he reminded himself he'd be meeting Mike Shapiro, Mike Shapiro who had become one of Bennett's heroes with remarkable speed. From the podcast, he had a sense of who Mike was as a person; he was someone who found joy in the success of others. Bennett certainly didn't think Mike was the type to be cruel just for cruelty's sake. So if nothing else, at least flying to San Francisco would give him a chance to meet his hero.

He remembered, on *The Lions' Den*, when a contestant's pitch was turned down unanimously by all the judges, the show played a signature lion's roar. The roar was famous, featured on all the show's TV ads and reenacted by its fans—reenacted live by the studio audience and then later, when the show aired, in living rooms around the world. "Rawr!" had become the show's tagline—and Bennett's palms started sweating when he imagined that "Rawr!" being directed at him.

He took a deep breath and tried to get his anxiety under control. One thing that he liked about *The Lions' Den* was that the judges were savvy, successful businesspeople, and when they rejected an idea, it was always for a good reason. Even when they were cruel, the rejections offered a sort of lesson, the prospect of something valuable to learn. Every rejection pointed to ways that a pitch or a product could have been done better.

Bennett had half a day, on no sleep, to learn those lessons and turn them all into a perfect ninety-second pitch.

• • •

He arrived at the TV studio ahead of schedule. Where else would he go in a town where he knew no one? He'd spent the last four hours downing coffee like a fiend at a Starbucks, but he felt like he might as well head to the Den a bit early to get the lay of the land.

Or, as he now thought of it, to read the tape.

Once Bennett got to the studio, he signed in with a woman at the front desk.

"You're Bennett Gates?"

"That's me."

She eyed him suspiciously. "You're early."

"I am." From the way she narrowed her eyes, he wondered if they weren't going to let him in. "Is that okay?"

"No one in this industry is ever early," she said. "In the future, if you don't want to look green, don't show up two hours before your call time, K?"

She slid his name tag across the desk and jerked her thumb toward the elevator. "Off you go."

On his way to the elevator, he looked more closely at his name tag. Where it was supposed to list his job title, it said *Talent*.

"Because I'm talented?" he joked with the security guard.

"Yeah, that's absolutely why," the guard answered, stone-faced. "Head to the greenroom, down that hall to the right."

Bennett was 0–2. He had not managed to charm either of the employees so far, which didn't bode well for the Lions. If this were a test in reading the tape, he was failing it.

The greenroom wasn't green, it was more of a cream-color off-white. It was an L-shaped room with a few sofas and a mini kitchen. There were already a dozen people inside, gathered in small groups in different parts of the room. *More "talent,"* Bennett guessed. His competition. Some of them were wearing hoodies, and some were wearing suits, but all of them looked deadly serious—pacing, rehearsing their pitch speeches in a whisper, staring intently into their laptops. It reminded Bennett of his college library during finals week: quiet and unbelievably tense.

Every one of these people was about to be fed to the Lions.

He felt his own anxiety ratchet up. Instinctually he reached for his phone and his AirPods. Considering he was about to see Mike in real life, he might as well do a little pregaming with prerecorded Mike to lower his heart rate.

Bennett found an empty chair in the corner. He angled his body away from the other jittery contestants and began to listen.

Welcome back to *Read the Tape*. I've got three little words for you: *Know your nut*. Today we're going to talk about managing your cash flow and paying attention to financials. You might think, *I'll worry about that later, Mike*. But if you keep pushing it off to some date in the future, then you won't have a strong foundation to build on, which means your inner genius may never get a chance to spark.

Capital is the fuel for every entrepreneur. It fuels the fire for whatever it is you do, which is why talking about capital is a necessary matter. If you want to be successful, then you've got to ask these questions: What is it that you're trying to pursue? What are you spending? What is your burn rate?

> You've got to ask these questions: What is it that you're trying to pursue? What are you spending? What is your burn rate?

I'll tell you a story. A few years ago, I was working with an innovative healthcare company founded by an icon in the healthcare industry. They had this perception that capital was not relevant to the product, which was bizarre to me. They made decisions like there was an endless supply of capital without yet having any sales. I was an investor and on the board of this company, so I had access to the financials. I looked at the burn rates and went, "We don't have the product yet. Let's cool it on x," where x was whatever thing they wanted to spend a lot of money on. Their attitude was "We'll figure it out," and then everyone would throw money at them. And they weren't wrong; people did throw large sums of money at them, because what they were bringing to the healthcare space hadn't been done before. But then the investors eventually want to start seeing a product and understand how it will be monetized. Luckily, as time went on, I was able to get them to think about capital differently in at least some ways, like cutting costs.

Don't get me wrong. The CEO of this company was brilliant. She had more than twenty years of experience running hospitals before she'd decided to take an entrepreneurial approach to reimagining healthcare. She was a genius. And thanks to all her years on the boards of some of the country's most prestigious hospitals, her credibility and capability were through the roof. That's why people threw money at this company: because of who she was and where she came from. This woman was light-years ahead of

me as far as capability, intelligence, and education—
yet I was significantly better in sales and running a
company. We were an odd couple, but that illustrates
another Mikeism: surround yourself with people who
have skill sets you don't.

This is part of knowing your nut, sure, but it's also
about being successful in whatever entrepreneurial
endeavor you want to launch. It's important for life in
general. I always have a clear calculation in my head
of what my financial statements look like, but I also
have a strong CFO with me at all times because I'm
also a spender. I like to roll out the red carpet, enjoy
good food and wine, and have nice things. Spending
has worked for me, but I've always had somebody
around me who's *not* a spender and who actually
starts yelling if they feel like I'm getting out of control.
And I listen. It's mandatory to surround yourself with
opposites, people with different skill sets than you
who can provide the facts and the intelligence you
need to make smart financial decisions.

Bennett felt his throat go dry. Smart financial decisions. What
were those, again? He thought of the money Uncle Jim had loaned
him. The money he had no idea how he was going to pay back if
things on *The Lions' Den* didn't pan out.

The whole idea of knowing his nut made Bennett nervous. In
his prior jobs, he'd never had to pay attention to the financials at all.
His college buds weren't too concerned about the money part—which
had ended up being a problem—and the brothers in the video game
start-up had smartly put a CFO in place, even when their team was

lean. Bennett neither had the right skill set to manage his cash flow nor sought out a business partner who did.

In a mild panic, he started scrolling through his iPhone, trying to find a different episode of *Read the Tape* to listen to. His eyes caught on something about a ghost, which sounded about as far from financials as he could get. Perfect. He hit play.

> There's always information around us. If you have the facts and the intelligence you need, you can always read the tape, negotiate, and figure out what direction to go in. It's when you're missing the intelligence that things get tricky. For example, say you have a competitor, but they're not public. So you can't read their financial statements, and you don't know who the owner is because it's buried in an LLC. If you go into a meeting and you know all the facts and you have the intel, you can maneuver. But that gets a lot harder if you have a ghost in the room.
>
> What is a ghost in the room? It's something unknown. The pandemic was a ghost in the room. Trying to negotiate without knowing all the facts is a ghost in the room. It can be anything that's unforeseen, unexpected, or unknown. I once had a real estate company come to town, and we didn't know anything about them. We were going, "Wait. How can we negotiate? How can we keep our agents from going there when we don't have all the intel?" When they went public, everything finally came to light because we could read all their public statements and get that intel.

> If you have the intelligence, you can fight. But if you
> don't, it's a ghost in the room.

Bennett felt his eyes refocus on the room around him. He felt like something of a ghost himself, lurking in a chair in the corner, studiously avoiding eye contact with the other contestants. It wasn't that he was trying to be rude—he just didn't want to psych himself out. He was already too much in his head about meeting the Lions, especially Mike, and everyone else in this room looked just as anxious as he was.

Well, *almost* everyone.

In the corner of the room closest to Bennett, a slender woman with pink jeans and thick-framed glasses was kicked back on a sofa. She was sketching on a tablet, humming softly to herself while she drew.

She was dressed like someone with software money—colorful clothes that probably cost more than Bennett's entire wardrobe but still looked as comfortable as pajamas. Her long box braids were dyed a trendy hot pink. She also had the telltale accessories of the creative class: the Apple Watch, the AirPods, the tattoos. She looked cool. She looked like someone who belonged at a Silicon Valley Pitch Fest.

Unlike Bennett.

"Do I know you?" she asked. She had caught him looking.

"Sorry to stare. I'm just nervous. I'm Bennett, from Virginia."

She squinted her eyes and looked at him more closely. "Bennett from Virginia, I do know you! You're the King of Karaoke! From Ex Machina!"

Bennett cringed. "You saw that?"

"*Saw* it? I told my girlfriend it was one of the best things about the whole conference!" She laughed. "You really sang your heart out.

I mean, it's awful music—no offense—but you gave it your all, and I admire that." She held out her hand. "I'm Gem, from Santa Cruz."

"Like gemstone?"

"Nah. Supposed to be short for Gemini, thanks to my mom—I always tease her about being the first real Black hippie in America. But I said no thanks to Gemini, so it's just plain Gem for me."

They shook hands, and Bennett realized how good it felt to be recognized by someone here, so far from home, and to be recognized as a peer, one techie to another. It made him feel less out of place, less like a poseur.

"What are you pitching today?" she asked.

He took a breath to answer and … drew a blank. "It's an app," he finally managed.

"Well, I should hope so!" she laughed. "What's it do? Who's it for?"

Bennett couldn't think of anything to say. He got a terrible, sinking feeling. This Pitch Fest was a horrible idea, and doing it under-prepared and on almost no sleep was an even worse idea. He was going to get shredded out there.

But then he surprised himself. He reached into his pocket and pulled out his phone. "Let me show you," he offered.

Gem looked at his phone and squealed with delight. "Ohmygod, I love this!"

"I haven't even opened the app yet."

"No, this!" She pointed at the photo on his phone's wallpaper. It was from a road trip he had taken with Maurice. While the truck was parked, the dog had climbed into the driver's seat and put his paws on the wheels so it looked like he was steering.

"Yeah, he's great," Bennett said. "His name is Maurice."

"You named your truck?"

"What? No. I named my dog."

"I don't care about your dog! Bennett, that truck! What is that, a '72 Chevy?"

"I guess."

"You *guess*! That truck is a thing of beauty! And it's totally a '72. It's got the ribbon on the grille." She let out a low whistle. "Not every white boy in Virginia's got a truck like that, Bennett. Talk about privilege."

She winked, and it took him a moment to realize she was joking.

"But I digress," Gem said. "Please, tell me about your app."

Bennett took his phone back and readied himself for what would be his first—and probably only—rehearsal run of his upcoming pitch. "My app is called RobotFindr, and it is designed to find robots."

"Excellent!" Gem said, with a giant smile.

Bennett started explaining RobotFindr and how it worked, but when he pressed the button to launch the app, Gem frowned. "Where's the front end?" she asked.

He stared at her blankly. "What do you mean?"

"The part the user sees? The look and feel? The design?" She said, "I'm a designer, so the front end, the experience of the user, is always the first thing I look for."

Bennett frowned. "RobotFindr isn't really that kind of app," he said. "Most of it runs on the server, so it doesn't matter what the user sees. I'll make it look pretty later."

"Okay," she said. "But there's a lot more to design than 'looking pretty.'"

He tried to explain. "What matters is the code. Behind the scenes, the code is doing amazing stuff, parsing these ads. And it's totally extensible." *Extensible.* That was a word tech people liked to throw around, and he wanted to be sure to get it in his pitch, even

153

though he was pretty sure it just meant *extendable*. "It can be taught to search for anything, not just robots."

But he could see he had lost her. She had been interested in his app, and now she wasn't—all because she couldn't see past its basic, utilitarian design. Hadn't Scottie said something similar? Why did people care so much what the app looked like? Didn't they understand what it could *do*?

A stage manager with a headset crew came in. "You're up next," he said to Gem.

Gem turned back to Bennett and gave him a halfhearted smile. "It looks like you have a lot of really smart code in there," she said kindly. "Good luck with it." Then she let the stage manager lead her toward the stage.

He stared at his app.

He was going to get torn to pieces.

Rawr.

• • •

The next two hours were among the worst of Bennett's life. He kept trying to prepare the speech he would give to the Lions, but he couldn't concentrate. All he could see was the look on Gem's face.

Where's the front end?

Why hadn't he thought of this? No one goes onto *The Lions' Den* with a prototype! Why hadn't he taken the time to hire a designer to make the RobotFindr app look good?

Because he hadn't had time.

Because he didn't have money to spend on extras like a designer.

Because he believed in his bones that if the app was good, then investors would be able to see that, no matter what the design looked like.

And the app *was* good. It really was.

But just as he started to get his confidence back, just as he got back to the point of being able to think about the pitch he was going to make, Gem would pop back into his head.

The part the user sees? The look and feel? The design?

"It doesn't matter what the user sees." That's what he had told her. But he could see now; of course it mattered. If someone opened the app and got confused, or got scared off, or thought it was too complicated, then they wouldn't use it—and they would never get to see his code or how good it was. They would, ironically, only get 90 percent of the way there … and then toss it.

Bennett had been so focused on making the computer code work he had completely forgotten that the back end only works *if someone uses the app*. People would see this, and they wouldn't want to use the app.

He should have made something prettier, more inviting. More clear. More … designed.

What was he going to do? How was he supposed to—

Mike Shapiro's voice brought him up short.

"I'm not gonna be that person. I'm not gonna invest in something because it looks really good on TV and is good for my own narrative."

For a moment, Bennett thought his AirPods had switched back on. But then he remembered he wasn't wearing them.

And then he *saw* Mike Shapiro's face on the widescreen in the greenroom. Someone had just switched it on. It took Bennett a second to realize they were broadcasting live from the TV studio down the hall.

He gulped. Everyone in the room gulped in unison.

"For me to invest," Mike went on, "I'm looking for companies that have a working product and happy customers, because I need to understand how the company is being valued. I read financial statements like other people read a book. I need to see your offices and meet every single one of your team members. Do you support them? Do they feel respected? Are their opinions taken seriously? And what about your customers? Are they getting what they need? Selling and marketing should always be geared to the consumer. If they do not like what you're selling, it's not a good product. You have to ask yourself, Who is my audience—and what do they aspire to?"

Right into the camera, Mike said, "What I'm saying is I need to see you perform for a minimum of three months with your product and your team."

Then he looked away from the camera and set his gaze right on the Lions themselves.

> Selling and marketing should always be geared to the consumer. If they do not like what you're selling, it's not a good product.

"I'm here as a guest in the Den because I've watched you Lions for years. You're all successful business people—I don't doubt that. But I've also seen the way you play to the cameras, and that isn't me. I'm not gonna say yes to something that isn't ready, just for the ratings. I'm not here to make good TV. I'm here to make good decisions and to help the entrepreneurs, because what is most important to me, in addition to business, is simple: finding joy in the success of others."

Bennett's jaw dropped halfway to the floor. Had Mike really just mopped the floor with the Lions—on their *own show*? Talk about firing a shot across the bow.

"Bennett Gates?" It was the stage manager in the headset. "You're up next!"

• • •

Afterward, Bennett remembered his entire experience of being onstage at *The Lions' Den* as if it had been a dream. He described it as an "out-of-body experience," though while it was happening, it felt more like "near-death."

As soon as he walked onstage, his eyes were blinded by the bright stage lights, so he could barely make out the Lions, the five silhouetted show hosts sitting on their thrones, the five people who might invest in his app and change the course of his life forever—or might mock him on television and ruin all his dreams.

One of the silhouettes spoke. "What have you got for us today?"

The voice was one he had heard over countless hours in these recent weeks. Ever since Uncle Jim had introduced him to the *Read the Tape* podcast, Bennett had been listening to Mike Shapiro almost every day, while he was driving, or cooking his dinner, or working out, or winding down at the end of the night. Over these weeks, he had taken a lot of comfort from these podcasts: he had come to think of Mike as his mentor and coach, and he had adopted many of Mike's lessons and seen them make a difference in his life.

So hearing that same voice here on *The Lions' Den* set calmed him.

And then he heard Mike's voice in his mind, too—some of the lessons he had been absorbing over these weeks. *Don't stress about your weaknesses* was one thing Mike had said. *Don't be afraid.*

Bennett took a deep breath. *I've got this,* he thought.

And then he heard himself speak, as if someone else were doing it, as if someone else were in control of his body. "This app," he heard himself say to the Lions, "will save a lot of people a lot of time and money."

• • •

They rejected him.

Every single judge on *The Lions' Den* rejected him.

In the end, it wasn't because of the app's design at all. Bennett had been prepared for that, had already braced himself for that, and had maybe circumvented it a little by warning the Lions up front: "It's kind of ugly! That's why I need funding!"

Everyone in the audience had laughed at his joke, and he breathed a gigantic sigh of relief. He had survived.

When the laughter died down, Mike spoke gently into his microphone. "Let's talk about your financials."

His financials?

Bennett didn't have financials. He had been so nose-down creating the app, he hadn't given a single thought to creating the business.

Rawr.

• • •

Bennett stood outside the TV studio, looking out at the San Francisco Bay and feeling the city's famous wind on his face. He felt hollow inside. Whatever momentum he thought he'd been building, whatever hope had been growing in him, was gone.

Now he was just a guy alone in a city of strangers, wondering what he was even doing there, wondering what he was going to do next.

Had he really imagined he was going to waltz onto *The Lions' Den* after one week of work and walk away with the grand prize? Had he really believed he would get a major, life-changing investment when he didn't even "know his nut"? Had he really thought that his hero and mentor Mike Shapiro was going to stand up, greet him with a big handshake and a pat on the back, and say, "Welcome to the big leagues, kid!"?

Yes, he had.

What a joke.

He looked at the city whirling around him—people zipping by on scooters and bikes, streetcars rattling along their tracks. San Francisco was alive, and these people were flowing through its veins like blood, carrying the oxygen of ideas with them. He could feel it. He could taste it. It was the same energy he had felt himself just a few days ago: the feeling that anything was possible. Now it felt a million miles away.

"Hey! Bennett from Virginia!" It was Gem, now wearing some rose-gold sunglasses, her hot pink box braids twisted into a loose bun. "How did it go in there?"

Bennett did a halfhearted big cat impersonation. "Rawr."

She laughed. "Same with us! Rawr!"

Somehow this made Bennett feel better—not because Gem had lost but because misery loves company. She seemed to know what she was doing, and if she could take the rejection with such lighthearted good humor, then maybe he should try to do the same. "Sing it with me, Gem. Rawr!" he said, and the two of them started roaring together, first quietly, then full voice, until the people in the passing streetcar turned and stared and some of them even started roaring back. "*Rawr!*"

They'd roared until their voices couldn't take it anymore, and then he asked, "So what do we do now?"

Gem shrugged. "I have a gig, but it doesn't start for another few weeks. I figured I'd spend the time surfing."

"Huh," he said. "I wouldn't have taken you for a surfer."

She rolled her eyes. "No one ever thinks the queer Black girl in tech can also surf. But I'm here to tell you, Bennett from Virginia, here in California we can do it all."

"I wasn't thinking any of that," Bennett said. "Honestly, I was thinking I wish I knew how to surf. Because the truth is, I have no idea what I'm doing next."

He tried to imagine what it was like being Gem—living in Santa Cruz, going from one tech gig to another, surfing in her free time. She was living the dream. Meanwhile, the closest Bennett came to fitness was driving by the 24 Hour Fitness near his parents' house—the house where he still lived, as a grown man, driving his dad's hand-me-down truck.

His dad's hand-me-down truck.

This gave him an idea. It was probably a horrible idea, or maybe too good to be true, but just having the idea got his heart racing. *Hope.* Right there where he had left it.

He cleared his throat. "Um, okay, so, remember that stuff you said about my app's design?"

"I'm sorry I said that! I had no business—"

"You're totally right. If I'm going to get funding for my app, then I need it to focus more on the users, the experience of using the app. I need to imagine what this is like from their point of view, not just from the point of view of the code. You have a few free weeks, right? So I was wondering if I could hire you to work on it."

To his surprise, she looked interested, so he kept going. Now the hard part. "Trouble is," he said, "I can't pay you. I have no money. I mean, that's why I came to *The Lions' Den* in the first place. But I'm not asking you to work for free either."

She looked understandably skeptical. "You're going to say 'stock options,' aren't you?"

"Yes and no." Stock—equity in his nascent company—was one way he could compensate her for her time. If the app ever took off, then she'd be one of the first people to make money off of it. Bennett still felt the sting of being fired from the video game start-up before his stock options vested. Though, to be honest, that all seemed very faraway now. He realized he couldn't remember the last time he'd felt sorry for himself about making the "biggest mistake of his life." He simply hadn't had the time. And frankly, he'd been so excited about RobotFindr he finally understood how not excited he'd been about the multiplayer game. There was no inner genius there, no spark.

The problem with offering Gem stock options, of course, was that if RobotFindr didn't take off, then the stock would be as valuable as Monopoly money. And he understood now that investing in a design—in a team member who could focus on the things Bennett himself didn't naturally see—was worth actual money.

Money which he didn't have.

"If you design a look and feel for my app, help me redesign it with the user in mind, and throw in some general brand guidelines," he said slowly, "then I will pay you the lump sum of one '72 Chevy pickup truck."

He was gambling with his last asset on earth. If his bet paid off and the app started making money, then he would never even miss the truck. If it didn't, then he was going to need a plan, top to bottom. But one thing he remembered from Mike's podcasts: *Know your nut.* Bennett needed to acknowledge his own strengths and weaknesses, and where he needed help, he needed to invest in getting that help.

Hiring Gem was a good step in that direction. It was a risk he was willing to take.

For her part, she didn't even hesitate. "It's a deal!"

• • •

"Uncle Jim!" Bennett yelled into the phone. He was calling from the beach, and the wind was making it hard to hear.

"You in a tornado, son?" his uncle said. "Sounds like you're inside a twister."

"Sorry—I'm at the ocean, and it's a windy day."

"I see how it is. Mr. Hollywood has his big, fancy television debut, and now he's a California beach bum." Jim chuckled. "You know the Den has always been one of my favorite shows, Ben. The Lions don't pull their punches."

"You're tellin' me."

"And those Lions took a bite out of you."

Bennett sighed. "That's putting it mildly, Uncle Jim. It feels more like they tore me apart from limb to limb."

Jim took a moment to respond, like he was weighing his words thoughtfully. "I gotta say, I enjoyed seeing Mike Shapiro up there as a special guest, especially after all those hours listening to the podcast. He's just as good on TV as he is on the airwaves. Mike really can *rawr* with the best of them." He let out a low whistle. "And son, he wasn't wrong about your idea. It's a good one. It's got a lot of promise—even I can see that, and I'm no tech wiz. But it's not fully cooked."

"No, he wasn't wrong." Bennett took a breath. "That's why I'm calling. I wanted to let you know I'm staying here in California a little longer."

It had been Gem's idea that they work together side by side over the next few weeks. It would be easier for her to get up to speed that way and easier for him to adjust the app on the fly, according to her recommendations.

He admitted to her, a little shamefully, that he couldn't even begin to afford to stay in a hotel in San Francisco for a month, or even a week. "No worries," she said. "Just stay with us in Santa Cruz. Couch surfing is an internet start-up tradition!"

Uncle Jim was excited about Bennett's change in plans, and he happily agreed to keep taking care of Maurice. "I can tell he misses you. Poor little guy! But you've got more important things to do."

Bennett *had* a lot of important things to do. When he wasn't working on the app, he was working on everything else. He gave himself a crash course in business school, googling things like "how to write a business plan," and began building a document that he would be able to offer to any prospective investor, telling them clear as day what the app was, who it was for, how it would grow, how much it would cost, and, most importantly, how it would make money.

When the business plan and the new version of the app were ready, he would go out again looking for funding—and he knew exactly where he wanted to start.

Mike Shapiro.

YOU ARE ALWAYS SELLING

NEWPORT BEACH, CALIFORNIA

"So what do you think of California?" Gem said.

She asked the question on the beach, as she and Bennett sat on her striped beach towel, eating the freshest fish tacos he had ever tasted from a little beachside taco stand, drinking Mexican Coca-Cola, and watching the blue Pacific Ocean crash onto the shore.

What did Bennett think of California?

"You've got to be joking," he said.

She made a serious face. "Do I look like I'm joking, Bennett from Virginia?"

He lifted his Coke to hers in a toast. "I think *you*, Gem from California, may be the luckiest person alive."

They were at the halfway mark of their road trip from Santa Cruz. There was a shorter path to Newport Beach, Gem had explained: "In a perfect world, it's only five hours if we take the 101 and the 5, but with traffic it's more like seven." He was learning that Californians really liked to talk about their highways, and which ones were better or worse for beating traffic.

"However," Gem had said, "nobody in the history of America has ever had anything good to say about the 101 or the 5. *I* recommend the scenic route. Let's take PCH," meaning the Pacific Coast Highway, which snaked down the coast. "It's more like ten hours, but it's worth it—even if just for the sea lions."

Bennett had no idea what sea lions were—but he'd said yes. He was getting better at saying yes. He thought of something Mike said on one of those early podcast episodes: *I want you to do something that makes you uncomfortable, something disruptive. Not something illegal, of course, but something you've been afraid to do even though you know, deep down, that it's really not that big of a deal.*

Like a road trip down the coast, Bennett thought. Like saying yes when Gem asked if he wanted to learn to surf, even though he was terrified. (He actually wasn't half bad—and after two weeks of catching the waves, he'd never felt fitter.) Like going to see Mike Shapiro, even though Mike had said "no thanks" to Bennett's big idea—on national television.

Bennett had failed—miserably, publicly. But if he had learned one thing from Mike—and, of course, he'd learned way more than that—it was to embrace failure, to *celebrate* failure. Because it was through failure that you found your potential and possibilities. If you could pivot from failure, get up, and move forward, well, that was the road to success.

A road that sometimes included a brief detour for fish tacos, elephant seals, and the most gorgeous ocean vistas Bennett had seen in his life.

A road that, not incidentally, led straight to Mike Shapiro's office door.

When Bennett had called Mike a few days earlier to ask for a second chance, he wasn't sure what to expect. But of course he hoped Mike really meant what he says about taking every call and every meeting, as he was dialing the phone. It seemed entirely possible that Mike Shapiro—businessman, investor, and podcast king, the guy who had become Bennett's de facto coach and mentor over the last couple of months—would say "You're the kid with the robot app?" and then hang up the phone.

Instead, when Bennett explained who he was and what he wanted, Mike didn't hesitate. He said, "Sure. I'm in Newport Beach. Come on down."

Bennett hadn't even felt nervous, only excited, because this time he had done his homework.

The last month had been the most intense, purposeful, and rewarding of Bennett's life. He had shifted his focus to the consumer in a big way. RobotFindr was no longer about his code or his ego or his brilliant idea. Again he heard Mike's words: *Selling and marketing should always be geared to the consumer. If they do not like what you're selling, it's not a good product. Ask yourself: Who is my audience—and what do they aspire to?*

When he thought of his audience, he thought of Scottie. He couldn't wait to show her how the app looked now. Gem really was a genius; she'd taken his clunky code and turned RobotFindr into a thing of beauty. It was ready for the big time.

Bennett realized—in a way that didn't entirely make sense to him—that he missed Scottie. They'd talked a few times on the phone while he was in California, though he was honestly so busy with the app he hadn't had a lot of free time. How could he miss someone he barely knew?

But soon enough, he told himself, he'd be back in Virginia, hopefully with the backing of a major investor. Because that was the other thing that had changed since his ill-fated television debut: he knew his nut. He'd spent the last three weeks doing a deep dive into the financials of RobotFindr. Finally, he was paying attention to the right things. He knew exactly how he planned to incentivize investors, how to compensate his team, and how to manage cash flow.

Now he just needed to get some cash flowing.

Gem stood up. "Better get back on the road," she said, brushing the sand off her jeans, "if we're going to see those seals."

"Wanna *Read the Tape?*" Bennett asked, as he folded up the beach towel. He'd been listening to the podcast religiously over the last few weeks, and Gem had really gotten into it. Sometimes her girlfriend joined in, and the three of them would hang out in the cozy Santa Cruz apartment, throw back a few beers, and "party" with Mike Shapiro, as they'd come to think of it.

"You mean, Do I wanna listen to my man Mike?" Gem grinned. "You know I do."

· · ·

How about a Mikeism that's evergreen? *You are always selling.* Whether business or personal, that's the truth. That's why you have to look the part of the people you're trying to sell to. It's part of your credibility. When I was selling real estate in Arizona, it was important to look the part, with nice clothes and shoes. In Seattle, it's the exact opposite: I've met with billionaires who have holes in their T-shirts; it's a race to the bottom. Compare that to Southern California, where there's so much focus on how you look that it can start to get conflated with who you are.

I'll tell you a funny story. A ton of real estate agents look like models—they literally look like they just stepped off a runway—which makes it easy to do a fashion show. So one time, back when I owned the real estate brand, the person running one of those fashion shows said to me, "Hey, Mike, would you be in the fashion show?" I started laughing hysterically, because obviously I am not a model. I'm shaped like a box, you know? But this person was serious. They said, "Can you please go in for a fitting?"

I'm like, "Don't worry about it. I'm just extra, extra, extra large."

I show up the day of the fashion show. Everyone else is like seventeen feet tall. The men's designer starts putting stuff on me. I'm like, *Oh my God, I look like I'm from the Jersey Shore. I'm wearing Baby Gap.* And I say to him, "I can't go out like this. Yes, I'd like to be a bodybuilder, but I'm not gonna show that at my age."

He got panicky, then said, "You know what? Just wear your own clothes."

So I notice there's a tray of food for the models. Then I see there's a DJ. So I run out, give the DJ a hundred bucks, and say, "Play 'Hava Nagila.'" And then I brought some food. So I'm eating, walking the runway in my own schlubby clothes, and they're playing "Hava Nagila." Everyone went wild. They were dying. The audience realized what was happening. Like ... I'm not gonna fit in those clothes.

Long story longer, the next year they raised all this money, and they said, "Mike, will you do it again?"

I'm like, "You've gotta be kidding me."

But they were serious. So I get assigned this kid who works for this custom suit guy who does a lot of clothing for famous people. The young guy working for him was maybe twenty-six—a good-looking guy who looked the part. So he came in and said, "Mike, this is my living. You cannot embarrass me; you can't ruin this. Please shut up and let me do this right."

So direct and powerful at twenty-six! I'm looking at this kid going, *Whoa! Okay, boss man!*

So I did it. I was appropriate. I got dressed in the clothes he gave me. Of course, all my friends and family were like, "Mike, you look amazing! Go buy clothes from this kid!"

He won the competition. He kept wanting to have lunch, and I kept having lunch with him. I kept saying, "Why are you doing this with someone else? Why don't you do something on your own? Go ahead. I'll help you out."

So sure enough, he quit the job working for the other designer. He came to me and said, "What do I do now?"

And I said, "No problem." I had him set up shop as a clothing company at my real estate company's headquarters, which meant there was actually a designer men's suit line at a global real estate brand's affiliate. All his customers thought it was the greatest thing ever.

It was hilarious, because my involvement in that fashion show started out as this big joke, and then in the end, I was able to help this talented guy. And it became very lucrative for him because his clients were so wealthy. Then some of them ended up buying houses, so the global brand liked it too.

That became another example of finding joy in the success of others. I asked myself, *How do I help this kid?* He was aware of what his capabilities were. The key to coaching him was teaching him to have a sense of humor. Humor is a very important part of my personality and the way I teach. Laughter, smiling, enjoying yourself—it's mandatory, because if you have people who are not enjoying their experience, it's over. Don't even bother. So I said to him, "You have to use humor when you sell to these guys. Otherwise it's not gonna work." Teaching him that behavior ended up being beneficial to him. He was scared to go out on his own—but once he did, he never looked back.

Bennett hit the pause button on his phone. He exchanged glances with Gem in the driver's seat.

"Are you thinking what I'm thinking?" he said.

"That I really wish I'd been at that fashion show to see Mike stuffing his face to 'Hava Nagila'?"

Bennett laughed. Gem had great comedic timing. Over the last few weeks, they'd laughed a lot together. Humor really did feel like a mandatory part of their relationship, and it made even the grueling days and long nights working on the app feel easy and enjoyable. In Gem, he didn't just have a business partner; he had a friend for life.

"What I was *going* to say," he said, "is that I really hope Mike helps us like he helped that designer-suit guy. I know he doesn't make bad investments—and to be fair, I think what I pitched him a few weeks ago *was* a bad investment. But I feel like I'm a different person now. I've been putting all the Mikeisms to use—and I've learned from my failures. I've never been more aware of my capabilities. And I'm sure as hell aware of yours! You're a genius, Gem, and I could never have done this without you. What we've been able to create together is epic. You know it is."

"Hold onto your designer suit there, big guy," she teased, but Bennett could tell she was happy.

"I feel like for maybe the first time in my life, I've actually found my inner genius. I want to build things that matter—to use my skills and talents to create real products that help people. And I honestly think this app has the power to do that."

"Well then," Gem said. "We better keep reading the tape."

> Let's talk about knowing your limits. Sure, I want to play in the NFL. But I am obviously not gonna play in the NFL. When I talk about knowing your limits,

it's about asking yourself, Is this an achievable capability?

A lot of coaches say, "You can be anything you want to be." No you can't. You have to take stock of what you're capable of. You have to take stock of your abilities. That's why I focus on inner genius. As a coach I ask, "What are you great at?" And if they say nothing, then I get them to go back to childhood. I say, "You've talked about this already. Was it easy to make friends? Were you great in math class?"

Everyone has an inner genius. The problem is that, for many of us as we get older, the genius gets clouded over, and we forget.

Knowing your limits is not supposed to be a barrier to success. It's more like: Why are you pursuing this when it's going to be the worst possible thing for you to do? Why make your life harder than easier? When I ask those questions, I'm just trying to maximize the result. I don't like a lot of the self-help bullshit along the lines of "Nothing is impossible." Or you'll read coaching books where everyone has all these big plans about what you should do and how to do it. There's this idea that, if you throw enough of whatever belief against the wall, something will stick.

I don't believe in that concept of coaching. More often than not, it leads you to places where you're not actually unlocking your inner genius. For example, people go into these multilevel marketing schemes all because someone tells them "You can be rich! You can do anything!" But the person saying that is this massively capable, charismatic person who's able to speak in front of large groups and doesn't care what people think. Maybe you *want* to be that person, but you're just not.

If that's what you want to pursue, fine. I'm not saying it's a bad thing. But know your limits. What are you good at? What are you capable of? What are you comfortable with? Do you want to be in front of a screen? What *can* you do? Knowing your limits is really about self-discovery. It's about personality and capability and figuring out how to make your inner genius shine.

Bennett hit pause again. "You know something? It's so helpful to think of things this way. A few months ago, I would have thought that 'knowing your limits' sounds negative, but really it's not. It's actually freeing. It keeps you from doing things like—oh, I don't know—pitching a product on national television when you have no idea what you're doing."

"Yeah," she said wryly. "It wasn't your best look."

"Hey now. We can't all be superhumanly confident."

She shrugged. "What can I say? I was born without the self-doubt gene."

Even though Bennett had known Gem less than a month, he believed it. He wasn't sure he had ever met a more confident person. He couldn't wait to introduce Gem to Scottie. The two women had a similar sense of irreverent humor, and he had a feeling they would get along famously.

"Look at that," Gem said, gesturing toward the ocean to their right. "Those waves are cranking. Too bad we don't have our boards strapped to the roof of the car." Her eyes sparkled. "Oh wait. We *do*."

Bennett grinned. He was eager to get to Mike's office and make his big pitch. But the truth was, he'd found that being out on the water, on Gem's old surfboard she had graciously lent him, helped clear his head.

Once again, he heard Mike's voice. *Keep moving. Don't get stuck. Move all the time. Go outside. Exercise your brain; exercise your body. It is absolutely mandatory. Moving keeps your head clear, reduces stress, and makes you more successful in every endeavor.*

Mike popped up in Bennett's head so frequently these days he might as well buy real estate.

"I really want this pitch to be successful," Bennett said, and saw Gem's shoulders deflate.

"Yeah," she sighed. "I get it. You want to stay focused."

"Absolutely." He grinned. "So let's catch some waves."

• • •

Surfing before the big meeting with Mike Shapiro was the right decision. When Gem pulled up outside Mike's office in Newport Beach, Bennett still felt a little nervous, but in a good way—a way that would keep him on his toes. Being out on the ocean had centered him, the stress ebbing away like the tide.

He could do this. He felt ready. He'd found his inner genius—now he just had to sell it.

You are always selling, Mike had said, which had never felt more true. *You have to look the part of the people you're trying to sell to.*

For once in his life, Bennett looked the part. He'd taken a cue from Mike's social media and donned a nice, black T-shirt and jeans. Nothing too fancy—Mike wasn't that kind of guy. The one place Bennett had splurged was on the shoes. Inspired by a story Mike had shared on the podcast, he had bought himself a pair of high-end Berluti shoes, well made with beautiful, high-quality leather.

Would Mike notice the shoes? Bennett had a feeling he would. After all, who was better than Mike Shapiro at reading the tape?

With Gem by his side, Bennett took a breath—and walked into the most important meeting of his life.

The minute he saw Mike was *also* wearing a black T-shirt, he felt immediately at ease.

"Bennett!" Mike said with a big grin, walking toward them. "Gem!" He shook both their hands warmly. "Great to meet you both. Or meet you again, without the TV cameras." He appraised Bennett. "Love the Berlutis. Nice touch. And you're a lot tanner than you were a few weeks ago. Have you been soaking up some California sun?"

Bennett felt himself blushing. Mike really did read the tape.

But here's the thing. *So did Bennett.* He'd learned enough about situational awareness to move beyond just thinking about himself. A few months ago—hell, even a few weeks ago—his brain would have been filled with questions like *How does Mike perceive me? What is he thinking about me?* He would have been trapped in only thinking about himself, rather than listening to others.

Instead, Bennett walked into the room and watched—observed, listened, learned. He absorbed the sounds, energies, smells. The late-afternoon sunlight poured through the windows in a soft-yellow haze. He heard a woman's voice coming from a back room, laughing. *Mike's business partner,* Bennett guessed, who was clearly on the phone. When he inhaled, he caught the aroma of food—many different kinds of food, if he wasn't mistaken. And when he appraised Mike, he noticed that his mentor looked even more buff and tan than he had on the show.

"My tan's got nothing on yours," Bennett said. "And for the record, I think you're built more like a bodybuilder than a box. But now that you mention it, I have been getting outside more. Gem here has been teaching me to surf."

"Love it. Some great waves between here and Santa Cruz." Mike turned to Gem. "How'd he do?"

She smirked. "We didn't do anything too heavy. But the kid's got promise. Not bad for a Virginian."

They all laughed, even Bennett, who didn't mind the good-natured ribbing—because, after all, a sense of humor was something they all shared.

"Who's hungry?" said Mike. We got you the whole menu from the spot down the street."

Bennett's jaw dropped. "No kidding?"

"Why would I kid about food? You've heard the podcast. Always make 'em eat. Everyone gets so excited about food that they drop their guard—and then I read the tape!"

Bennett nodded toward the back room, where the woman's voice had gone silent. She'd gotten off the phone. "Does your partner want to join us?"

"Good ear," Mike said, impressed. "That's Tessie, my cousin, who coproduces the podcast. I'm sure she'd love to join us for a bite."

Mike got Tessie, then made a round of introductions, then led them all back to the food table, where Bennett's jaw dropped even further. His nose had not deceived him; Mike really had ordered the whole menu. The food was delicious, and as they all dug in, the conversation flowed easily, especially once they started talking about

The Lions' Den. As much as Bennett hadn't cared for the other Lions, it turned out Mike wasn't a big fan either.

"I just don't get those guys," Mike said. "They're not bad people, and most of them have been very successful. But what the hell are they doing?"

"Mike," Tessie chastised. "Do you really need to fire another shot across the bow? Didn't you get that out of your system on television?"

"But I think you're right," Bennett said. "They're not looking for sustainable products. They're throwing money at whatever looks glitzy under the stage lights."

"Yeah," Gem chimed in, "it doesn't seem like 'smart investments' is a top priority for Hollywood."

"Or even doing good business," Bennett added.

Bennett had to admit they really were talking much more candidly than they would have if they'd been sitting around a table in a buttoned-up office. Food *did* make people drop their guard. Yet another Mikeism you could take to the bank.

He hoped that soon he'd have something a little more solid to take to the bank.

"The real problem with the Lions," Mike said, "is that every single one of them fails the twenty-four-hour test. It's not even their fault. That's just the nature of the show."

Bennett and Gem exchanged blank looks, causing Mike to ask, incredulous, "Neither of you knows what the twenty-four-hour test is?"

"Mike," Tessie sighed, "they haven't gotten to that episode of the podcast yet. We only released it last week."

Bennett grinned, enjoying the banter between the two cousins. He got a kick out of the way their energies balanced each other and how deftly Tessie managed Mike.

"The twenty-four-hour test," Mike said, leaning back in his chair, "is exactly what it sounds like. Most people react instantaneously to anything, whether it's good, bad, or somewhere in between. They just react. And sometimes it's a major decision. This is a classic example of communication today, because communication is so instant. So if you, say, get really upset, you might write an email or a text instead of sitting on it and not reacting immediately."

"Guilty," Gem said, sheepish. "The first start-up I worked at, I wrote an email like that to my boss. Believe me, he deserved it."

"Did you still have a job the next day?" Mike asked.

"Erm … no," she admitted.

"The twenty-four-hour test is always important. When we don't react immediately, we make better decisions, rather than just reacting to what's been thrown in our path. After twenty-four hours, more likely than not, you'll calm down and think through the problem. There are times I've been in the midst of a negotiation or a major decision and something just doesn't feel right. I'll say that out loud: 'My gut instinct is that it doesn't feel right. Let's give it twenty-four hours.' Of course, keep in mind this is for major decisions. Obviously you can't always take twenty-four hours. If you have to make decisions sooner, then it is what it is. You follow your gut, not what your heart or emotion is saying. What is your gut telling you? Gut is usually right. They've proven this. They've actually done studies on gut reactions, and they've shown that they're right more often than they're wrong. Which is interesting."

> When we don't react immediately, we make better decisions, rather than just reacting to what's been thrown in our path.

"But how do you learn to trust your gut?" Bennett asked, genuinely curious. "What if it doesn't come naturally?"

"It's learned behavior. These days, I usually don't need a full twenty-four hours. I already know when I'm reacting to something and my reaction is wrong."

"There are still times you react," Tessie said, arching her eyebrow.

"Sure. Sometimes I just react, even when I know it's wrong. And then twenty-four hours later, I'm like, *That was stupid*—at which point, I try to be honest with myself, to say 'I blew it.' That's not a weakness; it's a strength."

"It's true." Tessie nodded her head. "When Mike messes up, he always admits it. That's kind of his thing. And to circle back to another Mikeism, he really does use food to his advantage. Whenever he has meetings with someone we're going to be negotiating with, he always wants it to be at our office with food. Everyone's more comfortable that way. Food loosens people up. And if meeting at our location, we can control the direction better."

"It never doesn't put a smile on someone's face," Mike added. "It never doesn't break the ice. It never doesn't allow for more of a personal connection. So for me, it's effective."

"For me too," said Gem, who could eat anything she wanted and never put on weight.

"I'll tell you a story," Mike said. "You know by now that's what I like to do. My father was a storyteller. He's probably where I got it from. So, years ago, we had a meeting with this guy where I ended up not hiring him. By the end of the meeting, we knew he wasn't the right fit."

"And yet," Tessie said, "Mike still sent him home with a steak anyway. The guy was thrilled."

"When you do that, they're going to go home with a positive feeling about you, even if the business partnership isn't there."

Bennett swallowed. Was this Mike's way of saying he wasn't going to say yes to Bennett's pitch?

"Relax," Mike said, reading his mind. "There's no hidden meaning here. I'm just sharing a Mikeism. Sometimes food is a good way to distract people. Like if I'm negotiating, someone will say something, and I'll respond with 'You look hungry! Wanna get some food?' It's not manipulation; it's just a tool. At the same time I'm putting the hammer down, I'm also making a nice gesture. That day, even if I didn't hire the kid, he still went home with a nice steak."

"Well," Bennett said, marshaling all his courage. "I hope I go home today with more than a nice steak."

Bennett glanced at Gem, who gave him an encouraging nod. It was now or never. He took a deep breath.

"Mike, I listened to every single word you said on *The Lions' Den*. In fact, I've watched the episode over a dozen times, even though it means watching my own humiliation. Actually, let me reword that: watching my own *failure*. Because, thanks to you, I think about failure differently now. Failure isn't just a part of being human; it's the most powerful weapon we have."

Mike leaned forward. "Go on," he said. "I'm listening."

"I want to tell you what I've learned, what I've done, and what I'm going to do. But first, I want to tell you about what your podcast has meant to me. Because you, Mike Shapiro, have changed my life."

And there, in Mike's office, after months of listening to Mike's stories, Bennett told a story of his own.

He told Mike about how he'd found his chutzma and the journey he'd been on the last few months to know and enjoy himself.

He told Mike about his inner genius—how being honest about the things he was *not* had helped him uncover the things he *was*.

He told Mike about the ways he'd learned to read the tape, getting out of his own head and his own insecurities to become aware of other people and listen to them instead.

He told Mike about how he'd read everything he could get his hands on, and learned to surf, and even about the woman he'd met who loved robots, who had sparked the greatest idea he'd ever had.

Bennett told him about RobotFindr.

But this was not the scattered, overidealized, narcissistic version he'd expressed (expressed poorly, he might add) on *The Lions' Den*. This was something altogether different.

"When I was on the *Den*, I had the dream, but I hadn't put in the time or the effort—not enough, anyway. You always say to 'Get going,' and I'd definitely gotten started—but it was still a hobby. Now it's so much bigger than that.

"RobotFindr is going to help a lot of people. It's technology that makes life easier for people. Because that's who I am, Mike. I'm a coder, but just doing code for code's sake—just going through the motions for someone else's app or video game—never lit me up inside. No. My inner genius is *finding ways to make a product that will help others*. To create tech—a robot, you might say—that can free up more time and more resources so that the humans get to enjoy more freedom. Because at the end of the day, that's all anyone really wants, isn't it? Freedom to do what they want to do and the confidence to go and do it."

Bennett could tell he had Mike's full attention. Reading the tape was easy; everyone in the room was rapt. They were hanging on every word.

"Now let's talk figures," Bennett said. "Mike, I know you said you're looking to invest in products that are actually good products and have sustainability. RobotFindr checks both those boxes, and I'm happy to share the business plan I've spent the last month putting together. I know you said that, before you invest in a company, you need to meet every single one of their team members."

He gestured wryly at himself and Gem. "You're lookin' at 'em. Someday we'll have more, but for now it's just the two of us. Gem is the best damn app designer I've ever had the pleasure to work with, and I will sing her praises till kingdom come. I swear on my grandfather's grave I will spend every day we're building this product listening to Gem, and learning from her, and respecting her judgment. I take her opinion seriously, and always will.

"I've heard you talk about how, ultimately, investments are in people; they're not necessarily in the product. So I'm asking you to invest in RobotFindr, sure. But even more than that, I'm asking you to invest in me."

But Bennett wasn't finished.

"I know you won't make an investment for three months. That's all I'm asking for. Let Gem and I launch RobotFindr, then give us three months—three months to watch this product perform, to watch *us* perform. At the end, you'll have three months of financial statements, and you can decide whether or not to invest."

That was it. That was the pitch. No matter what Mike said, Bennett knew he'd never pitched anything better in his life. He had come to Newport Beach and said everything he needed to say, everything but one thing.

He geared up for the final crescendo.

"Thanks to your podcast, I feel like I've spent the last few months hanging out with you, Mike. Even better, I feel like you've been my personal coach. I've learned more from you than I could ever express, and no matter how many times I say 'thank you,' it's nowhere near enough. I know you find joy in the success of others. Right now, I'm asking you to help me find my own success."

The room was completely silent. Bennett realized he was holding his breath. Everyone was.

If he was expecting a big speech in return, he didn't get one. Mike said one word.

"Yes."

•   •   •

Bennett's red-eye flight back home touched down on the tarmac just as the sun was coming up on a new day. He watched it as it turned the distant mountains purple, then pink, then warm orange. It was a new day.

As he grabbed his bag and headed off the plane, he felt lighter than he had in years. This trip to California had gone nothing like he'd expected it to. Instead, it had turned into something else altogether: the best opportunity of his life.

Bennett hadn't reached the 90 percent only to turn back around, like he had every other time in the past. Instead, he had kept moving, barreling all the way to 100 percent. And when he got to the wall, he hadn't just touched it. He'd pushed the hell out of that wall to see how much farther it could go.

He had met Gem, who had designed a beautiful, brilliant interface for the app. Even more importantly, she taught Bennett to think less about what he wanted his app to do and more about what the app's users would want it to do. This was a crucial difference, and the difference showed in his recent builds of RobotFindr. Before Bennett's trip, his app had been little more than a bundle of behind-the-scenes computer code. Now, only a few weeks later, it was a complete user experience.

Most important of all, he had met Mike Shapiro—twice!

The first time, he'd been rejected.

The second time, he had not.

Mike had been downright encouraging about RobotFindr. He seemed genuinely excited to invest in three months—assuming Bennett could make good on his promise. Together they had gone over expectations, accountability, and profit motive. "I don't doubt you'll get this off the ground," Mike said, "and I'm rarely wrong about who I believe in." He'd even given Bennett a little money to tide him over. "A safety net," he said, "with more to come."

The biggest difference was that now Bennett believed in himself.

He had been so fixated on writing good, smart software he hadn't given much thought to how that software was supposed to make money—for him or for his investors. His goal had been to create something that could solve Scottie's specific problem: figuring out where to place internet ads. And he had done it. He had created a software "robot" that could crawl the internet and find the best places for her to advertise her product.

He knew that a tool like this would be valuable to people—maybe very valuable—but up until now, he had given most of his attention to just proving it could work.

Now that was done, and he needed a business model.

Uncle Jim had offered to give Bennett a ride from the airport, and he'd even brought Maurice along. The dog squealed and squirmed and yelped when he saw Bennett, and Bennett couldn't help but smile as the dog ran into his arms.

"It's good to be loved, isn't it?" Uncle Jim said.

On the drive from the airport, Bennett told his uncle about everything that had happened in California and everything he had learned. Jim was a great audience for the story, nodding enthusiastically as Bennett described his meeting with Mike Shapiro in Newport Beach.

"So what are you thinking for your business model?" Jim asked. "How are you going to turn this into a moneymaker?"

Bennett had been thinking about this ever since the meeting with Mike. "Corporations spend a lot of money trying to figure out the best ways to spend their advertising money. They hire outside professionals and keep all sorts of metrics, but, end of the day, they're just guessing. They're spending millions of dollars on guessing."

"I guess that's true," Jim said wryly.

"But RobotFindr uses artificial intelligence—uses an actual robot!—to find the places where ads are most effective."

Jim laughed. "That's why you call it RobotFindr? This whole time, I thought it was because it searches the internet for robots."

"It searches the internet for anything we tell it to. And when it makes decisions about the best places to buy an ad, it's not guessing at all. It's doing what only a computer can do: it's finding patterns that humans can't see."

"Brilliant," Jim said admiringly.

"So," Bennett continued. "A piece of software that can save a company millions of dollars must be worth millions of dollars, right?"

Over breakfast, Bennett explained what he had in mind for a business model. The way he saw it, there were two different revenue streams. First, any company who wanted to use the software would pay a licensing fee to use it. Bennett didn't know yet how much he should charge, but he knew he could sell a basic version and also offer a customized enterprise version, for a big client with a special need.

He could probably build an entire business out of customizing Robot-Findr for every *Fortune* 500 company—and as long as his software cost them less than whatever they were spending now, they would be happy to pay him.

"You said there were two revenue streams," Jim said. "So what's the other one?"

"Well," Bennett said between bites of his omelet, "every time RobotFindr identifies a place to put an ad, the client will buy an ad for that page, right? That means my software is going to provide a valuable service for the people who are trying to sell ad space. It is going to help them sell a lot of ads. So I'll take a commission on the sale of the ads."

Uncle Jim was genuinely impressed. "You've come a long way!"

"I guess I have, Uncle Jim. I mean, I'm sort of making this up as I go."

"Yup. That's business, kid."

Bennett sighed. "The truth, Uncle Jim, a lot of the details are still hazy. How do I find clients? How do I negotiate contracts?"

"You know what Mike always says: sometimes the best negotiating skill is walking away. You want to get people to like you and interact with you on a personal basis, hence the food bit. If you make it personal, they tend to be more likely to move directionally. But you should always be in the mindset to walk away."

"You've had a lot more experience negotiating than I have, Uncle Jim. That stuff's like a foreign language to me. I'm the guy who codes, you know?"

"Sounds like maybe you need a partner—somebody who maybe isn't a software genius like you but who knows a thing or two about how to sell." Jim raised an eyebrow. "Somebody like me."

Bennett was shocked. "You told me you weren't going to help me with this."

"And I'm not! I'm not offering to 'help' you. I'm saying you just pitched me a really solid idea, and as a businessman, I want in."

Bennett looked across the table at his uncle. He realized, ever since this crazy adventure had started, he had felt like a little kid around Uncle Jim, always playing catch-up, asking for favors, trying to earn his uncle's respect. Now, he could see he had his uncle's respect, and instead of feeling like a kid, he felt like a peer.

"You can be my partner," Bennett said. "In fact, I'd like that very much."

"Glad to hear it." Jim grinned. "In honor of Mike, we'll start with food. Breakfast is on me."

• • •

The weeks after launch went by so fast Bennett could barely account for them. He had never been so busy in his life, taking meeting after meeting with Uncle Jim and what seemed like an endless list of possible clients. In the beginning, during those first few meetings, Bennett felt like he was playacting the role of the head of a tech company, despite the fact that it was actually true. But then, imperceptibly, the meetings became second nature to him. The fact was, he loved talking about RobotFindr. He loved talking about it with the CMOs as much as he loved talking about it with the CTOs. He loved answering their questions, and he loved seeing their faces as they realized how his software would work for them and how much time and money it would save them.

"Take every meeting," Mike liked to say. "Take every call. You're always one person away from the best idea of your existence, always one phone call away from whatever success you're trying to achieve."

Bennett was living that Mikeism every day, especially once people started saying yes. He couldn't believe it at first. Having an idea of a company was one thing. Having to serve paying clients? That was something altogether different. Sometimes he still slipped back into his old habits of self-doubt. "I can't do this, Jim! What if I mess up? What if it doesn't work? What if ..."

But Jim was always ready to talk him down. "Then we'll make it up as we go."

> You're always one person away from the best idea of your existence, always one phone call away from whatever success you're trying to achieve.

The time he wasn't spending in sales meetings, he was either spending writing custom code for these new clients or on video chats with Gem, who was still working with him to make the app even better after the soft launch. Despite Bennett's dreams of renting a big open-floor-plan office, he and Jim had agreed that there wasn't any need for the extra expense of office space. Not yet. They didn't have any other employees, after all.

But now that they had actual users, Bennett and Gem had to be more attentive than ever to that "user experience." And boy, did they have users. The number of downloads doubled, then quadrupled, then skyrocketed so rapidly he could no longer keep up.

Time moved at a breakneck pace, like a wave crashing into the shore. At the end of three months, Bennett had passionate, engaged users. He had big-fish clients from *Fortune* 500 companies. He had

189

outstanding financial statements. He had a brilliant, consumer-friendly app that was doing everything he had promised it would—and more.

The best thing about Mike calling to say he'd be investing a sizable sum in RobotFindr was that Bennett wasn't even surprised.

"Creativity is key," Mike said. "You get paid for ideas, and I've been blown away by what you've created. I'm 100 percent in. Now stop writing code, and go celebrate with someone special!"

Bennett knew immediately who that someone was.

· · ·

He had hardly seen Scottie for more than a quick coffee over the past three months. She'd been out of town, visiting friends and then going to two different weddings, and he'd been so busy he had almost forgotten how much he enjoyed her company.

Almost, but not quite.

A knock on the door. As he looked up to see Scottie, a warm feeling washed over him. She was standing in the doorway, holding—

"I brought you a robot," she said, smiling.

She had built him a custom robot sculpture, a one-of-a-kind piece of art.

"I love it!" he said. "A company mascot."

"Yes!" Scottie agreed. "And when you have a big, sprawling campus in Silicon Valley, I'll come out and build a giant version you can put at the entrance."

He poured them lemonades, and they caught up. Her art sales had been going through the roof, thanks to RobotFindr, and she had been spending a lot of time at her foundry, building new sculptures.

"So much more fun than spending time at my computer placing internet ads!"

Bennett didn't know how to describe the way he felt, sitting there with her, after getting what was probably the best news of his life. Not all the code in the world could capture the feeling. He wasn't sure he'd want it to.

And then he realized, with a sudden burst of surprise, what he was feeling. He thought of all the feedback he'd gotten, from past bosses and colleagues and recruiters, the words that had crushed him time and time again. Only now, those words were inverted. He was bringing his whole self to the table with Scottie, no longer afraid of failing—and no longer afraid of succeeding either. He was all in.

"Hey, Scottie," he said, reaching for her hand. "What are you doing tonight?"

She laced her fingers into his, like it was the most natural thing in the world. "I dunno. Wanna sing some karaoke with me?"

Bennett said yes.

FINDING JOY IN THE SUCCESS OF OTHERS

NEWPORT BEACH, CALIFORNIA

"And that's when I realized," Bennett said, bringing his long story to a close, "that maybe I could be successful in more than just business. Maybe it was time to pay as much attention to my personal life as I'd been paying to my professional success."

"No more ninety percenter for you," said Mike with a grin.

Bennett grinned back. "I've blown so far past 90 percent I can't even remember what it looks like."

He was in Mike's recording studio, podcast headphones clamped over both ears, speaking into the microphone. Had he really just shared his entire story? How many hours had gone by? He felt like he'd been in the studio for days. Mike had a way of getting him to spill everything, to be way more fearless than Bennett thought of himself as being. Though considering everything he'd said yes to over the last year, he seriously needed to rethink this perception of Bennett Gates as someone who didn't make big choices or take big risks.

"I think it's important for anyone to share their lives," Mike said. "It's important to me in *my* life. That's not to say it's easy. I have a very successful marriage. We've been married thirty-five years, but my wife and I are very different—I'm the crazy Jewish guy, and she's an upright WASP. The juxtaposition of that is funny. I'm loud and obnoxious, and she's like, 'Oh my God.' But our lives have just been filled with joy and all the funny things that have happened. We met in college and had mutual friends. I felt good about myself as a young man, for no reason whatsoever, and Tara felt good about herself. One night we were at a mutual party, and she said to me, 'Who the F do you think you are?' And I was in love."

Bennett laughed so hard he accidentally banged into the mic. He heard the sound crackle in his headphones.

"Sorry!" he called out to Tessie in the next room.

"It's fine. We'll fix it in post," came Tessie's voice through his headphones. "And you don't have to yell, Bennett. Thanks to technology, I'm right here."

"See?" Mike said, chuckling. "I like strong people in my life—Tessie being one, my wife being another. She claims the reason she was attracted to me was because of my shoes, which is really funny, and really insulting. There's nothing else about me that you like besides my shoes? When she repeats that story with people, they die of laughter. But it's worked for us. Thirty-five years later, we are highly integrated into each other's existence. There's been a lot of humor and fun. Through my lens, it's an important part of life, having a significant other."

"I agree," said Bennett. "This one's for you, Scottie."

"That's very sweet. But let's go back to RobotFindr for a second. You haven't told our listeners what happened a month ago. You saved that juicy little tidbit for the end."

"Oh," said Bennett. "Right. Well, a month ago, we got an offer, a really *good* offer. We're at six employees now and expecting to add two more before the year is out. We're growing fast. So not a huge surprise that someone wanted to buy RobotFindr."

"Of course," Mike said.

"But we couldn't find much about this guy's financials or really any intel whatsoever. Talk about a ghost in the room."

"Mikeism number one."

"Yup. So we decided to have him come down to the office. We've got a great space now—open floor plan, lots of light. Did I mention we just added a small gym annex so my team can blow off steam during the day? Or I can. Not a lot of surfing in Virginia, so I've traded my board for weights. Anyway, this guy comes down to the office, and I order pretty much everything off the menu at my favorite restaurant. Have it all delivered so it's there waiting for him."

"Mikeism number two."

"Absolutely. So he comes in and is kind of acting squirrelly. Doesn't touch the food. Who looks at a table full of food and doesn't eat any of it? *That* was weird. I thought, *Wow, I've really failed at making conversation flow more easily and getting this guy's guard down!* Still, the terms of the offer seemed really good, so I didn't want to kibosh it sight unseen. But let's just say I walked into the negotiation ready to walk away."

"Mikeism number three."

"One hundred percent. When we asked for more info on his financials, he seemed gun shy. Something about it just didn't feel right. I read the tape. Mikeism number four, I know. I listened to this guy. I observed. He's a Silicon Valley guy, very slick, who launched an app a few years back that's got some similarities to ours. I sensed from several things he said that he either wanted to buy RobotFindr and then push me out, which was no good—I still want to be involved—or he just wanted to buy us to disassemble our team from the inside and take out the competition.

"Like I said, it just didn't feel right. So I deployed Mikeism number five: the twenty-four-hour test. I gave it twenty-four hours. And the next day, I felt confident in my decision. I said, 'Thanks but no thanks.'

"Turns out, my gut instinct was exactly right. Only a few days later, my team uncovered all sorts of shady stuff about this guy and his start-up. Let's just say his former employees were *not* big fans, and neither were his customers. His deal may have looked good on the outside, but it was rotten to the core."

"Go on," Mike said, dangling the bait. "What happened next?"

"We got another offer the very next week from a well-reputed investor who came to our office *and* ate our food. She's fantastic. We see eye to eye on most things, and the places we don't are a good opportunity for growth. She's got a terrific vision for RobotFindr, and best of all, I'll still have skin in the game, at least until I dream up my next big idea."

"And how much was the offer you accepted?"

Bennett leaned into the mic with a sly grin. "*Three times* the amount of the first guy."

"That explains the shiny, new car parked outside."

"It sure does."

Mike was grinning from ear to ear. "People are so fearful of admitting failure. But it's what this whole podcast is built on! Failure is *good*. So if you fail, announce it, rejoice in it, own it. Celebrate your failures, because each one gets you closer to success."

"You're preaching to the choir," Bennett said. "If I hadn't failed on national television, I wouldn't be here."

"Exactly. And as our listeners know, it's an important part of my story that I've failed many times. I've blown it. I've made epic disasters. I hate when really successful people whitewash things, when they stand up at a college commencement and tell the story of their 'golden years,' leaving out anything less-than-positive in their past. When I talk at colleges around the country, I start with all my failures. I talk about pretend playing the cello and swimming at the wrong end of the pool. The students always start laughing. Then they all crowd around me at the end of the speech, because I was real. I was human.

"That's why failure is important to share. Every single successful person has failed. There's no way they haven't. Failure is part of being human—and it's the most powerful weapon you have."

Bennett reflected on the many failures that had led him to where he now sat, in Newport Beach, recording a podcast with his coach, mentor, and hero. Too many to count. And every one of those failures was worthy of being celebrated.

"I can't thank you enough," Bennett said sincerely.

"You don't have to, Ben. I don't have kids. I love helping young people like you. It's one of the ways I stay relevant. What a lot of kids don't understand is that everything runs on a cycle. Think about all the great singers. Cher has had a hit in every decade. So has Madonna. How do they do that? They change their relevance. They change who they are and what they're doing in order to establish themselves in each decade. And then they become legends."

"Mike," Bennett teased. "Are you trying to tell me you're about to launch a career as a pop singer?"

"Not unless you think the world is ready for me to go back on that runway, stuffing my face and singing 'Hava Nagila.'"

They both erupted into laughter, until Tessie told them both to focus.

"For your sake," Mike said to Bennett, "I hope you never have to hear me sing anything, even karaoke. But the way I stay relevant—what fuels me—is opening doors for other people and giving them a platform they never would have been exposed to. Opening doors for other people is a huge part of why I started this podcast. And it's why I have no plans of retiring. I want to keep working and doing this podcast or whatever other ideas I come up with to stay relevant. It's so important to well-being, health, and happiness. I mean, look at Charlie Munger and Warren Buffett. They're still relevant in their late nineties."

> What fuels me—is opening doors for other people and giving them a platform they never would have been exposed to.

Mike leaned back in his chair. "I don't expect I'll ever stop finding joy in the success of others. Sometimes all you have to do is put your hand under someone and say, 'Here's the net.' That usually allows creativity to happen and for success to blossom very rapidly. I mean, look at you!"

Bennett found himself getting a little choked up. "You changed my life, Mike."

Mike was beaming. "For me, those are the greatest words I can hear."

With an effort, Bennett pulled himself together. Even if *Read the Tape* listeners couldn't see him blubbering, he didn't want Tessie to have to edit out his tender sobs in post.

"Well then," he said, clearing his throat. "Now I've got an offer for *you*, something I truly hope you'll say yes to since I'm only in California so often."

Mike leaned in, curious. "I'm all ears."

Bennett grinned. "Wanna catch some waves?"

APPENDIX A

MIKEISMS

There's no magic here. These are the simple steps you can take to success.

Below are Mikeisms that Bennett learned in this book—and now you, too, can put them to use in your life.

Read the tape. To "read the tape" refers to the old practice of using a telegraph to transmit stock price updates. But when I say "read the tape," I'm talking about something much bigger. It's about reading the room, reading the environment, and reading the momentum of what's going on around you. Reading the tape is really about situational awareness. And when you're doing that with other people, there are three steps to keep in mind:

Step 1: Be open and real—and let the other person lead.

Step 2: Find a way to relate.

Step 3: Ask the right questions.

Reading the tape will help you accurately assess people, situations, trends—and then use that information to do the right things at the right times in order to move forward.

Chutzma. Chutzma is a combination of two things: charisma and *chutzpah*, a Yiddish word for "self-confidence." When you bring self-confidence and charisma together, you start to develop a deep belief in yourself and your capabilities. Chutzma means playing to your strengths. Instead of stressing about your weaknesses, find out what you're good at, and strengthen your belief in yourself. Developing chutzma doesn't require you to change who you are, but rather to figure out how to bring out the best parts of yourself and use those to your advantage.

Find your inner genius. Everyone has a charismatic quality, whether they know it or not. I call this your "inner genius." Your inner genius is something that makes you special, sort of like your fingerprint. What is your X factor, the thing that has always felt easy and fun, as if you were born to do it? What is it that brings you joy? *That's* your inner genius. You must always be your authentic self and not focus on who or what you *think* you should be. You are always going to be your own best marketer and salesperson. If you find your inner genius, selling will be fun and natural.

The Maxine maneuver. Maxine was my Yorkie who weighed all of five pounds and fit in the palm of my hand—and yet had an amazing ability to bend people to her will. Whenever she needed attention or wanted a treat, she would perform what became known as the Maxine maneuver: she stopped whatever she was doing, flopped onto the floor, rolled on her back, and stuck out her tongue. Almost every single time she dropped into her pose, she would get what she wanted, instinctively using her small size to her advantage. Maxine found a way to channel her inner genius. Because of that, she had more charisma and confidence than many other big dogs I've met— and even some people. If a scrappy little dog can figure out how to use her apparent disadvantages to her benefit, you can too.

Celebrate failure. Not only *will* you fail—you *should* fail. Remember me pretending to play cello? Or my dad cheering me on when I swam on the wrong side of the pool? Today we hear "failure is not an option" all the time, but failure absolutely *is* an option. Celebrate every single time you fail. It's how you pivot within those failures that will define your success.

Read, read, read. Read books, read articles, read newspapers every day to know what is going on in the world and so you can talk to other people. Knowledge comes from exposure, and exposure is everything. That's what knowledge is. Exposure! So read, read, read. And read the tape to know what is going on around you.

Keep moving. Fitness is incredibly important to my daily experience. I encourage you to make it important to yours. Get outside. Walk. Jog. Whatever you do, don't get stuck. Just find your own way to move all the time. Exercise your brain; exercise your body. Moving keeps your head clear, reduces stress, and makes you more successful in every endeavor. Keep moving, pushing yourself, pushing your boundaries—and you just might get inspired to take on more "weight" in other areas of your life.

Everything comes back, but it comes back differently. Nothing is forever. Take any significant societal disruption, like a natural disaster, a pandemic, or war. Afterward, when life begins to return to "normal," we notice that things aren't quite how they used to be. Some changes are obvious, and some are more nuanced. But when you get good at reading the tape, you start to ask the right questions about how things will come back differently. It's those questions that help you move forward, whether it's by suggesting improvements or launching a big new idea. You cannot get stuck. You must be able to pivot to something different.

You only have to be right 51 percent of the time. Think about it. If you're wrong 49 percent of the time, so what? If you take big risks, even if you're only right 51 percent of the time, you're still going to come out on top. So why not leap when others don't and look where others won't? Risk is right!

Above all, try something. When you read the tape—knowing that everything comes back but comes back differently and that you only have to be right 51 percent of the time—you can spot whatever challenge or opportunity is coming around the curve and use your inner genius to get ahead of it. In the words of FDR, "It is common sense to take a method and try it. If it fails, admit it frankly and try another. But above all, try something."

Always know your nut. Your "nut" is the money part. If you want to be successful, you have to manage your cash flow and pay attention to financials. And if money isn't something you can manage well, hire someone who can. Surround yourself with people who have skill sets you don't.

Fighting the ghost in the room. A "ghost in the room" is something unknown. The pandemic was a ghost in the room. Trying to negotiate without knowing all the facts is a ghost in the room. It can be anything that's unforeseen, unexpected, or unknown. If you have the intelligence, you can fight. But if you don't, be prepared to fight the ghost in the room.

You are always selling. It is important to be aware in both your business and personal lives that you are always selling. You are your own best marketer and salesperson. You are your product or business and are the representation of your product or business. You never know who you may run across throughout the day, and you've got to look the part of the people you're trying to sell to. It's part of your credibility. Pay attention and stay on brand.

Deploy the twenty-four-hour test. Most people react immediately to anything, whether it's good, bad, or somewhere in between. They simply react, even when it's a major decision. But when we don't react immediately, we make better decisions, rather than just reacting to what's been thrown in our path. So whenever you can, take twenty-four hours. After twenty-four hours, more likely than not, you'll have calmed down and have clarity.

Always be consumer focused. Selling and marketing should always be geared to the consumer. If they do not like what you're selling, it's not a good product. You have to ask yourself, Who is my audience—and what do they aspire to?

Food is a tool. Food is one of the best tools you have for negotiating. Eating is an important part of bonding and connecting. It makes everything more human. Everyone gets so excited about food that they drop their guard—and then you can read the tape. When meeting with people you are negotiating with, try to set up the meeting at your location with lots of good food. Everyone's more comfortable that way. Food loosens people up. And if meeting at your location, you can better control the direction of the meeting and negotiations.

Sometimes the best negotiating skill is walking away. You want to get people to like you and interact with you as human beings—which is why food and eating is important. If you make it personal, they tend to be more likely to move directionally. But you should always be in the mindset to walk away. If you walk away after a great meal, they will have positive memories of working with you and most likely reach back out soon.

Take every call and every meeting. Always take every call and every meeting. You are always one person away from the best idea of your existence and always one phone call away from whatever success you're trying to achieve. You never know who you may meet, what idea may be sparked or created, or what opportunity may exist. Networking and meeting people from many backgrounds and experiences are more important than staying within your personal network.

Creativity is key. This one is very important and even more so in these new times with everything changing. You get paid for your creativity. You are in the best position if you have the ideas. Be fearless in expressing your ideas. It helps you look at the problem as your own. The important thing is that you are able to pivot in your own mind while being creative as things change.

Age does not matter. Nobody is ever "too old." It's always a good time to start something, to start over, to begin, to achieve. Be like Rupert Murdoch and Charlie Munger. They are in their nineties and still relevant, still creating, still making great business decisions. Stop worrying about time frames and if you should be at a certain place or have a certain amount of money. Don't rush your life. Enjoy it. Find the thing you're passionate about, and become excellent at it. Your success will come that way. It's all about finding your passion and the things you love the most.

BONUS MIKEISMS

These Mikeisms haven't been taught to Bennett yet, but they will be very soon. Here is a list of what is to come.

You can follow Mike at

> mikesshapiro.com
> [IG] @shapiromethod
> [LinkedIn] linkedin.com/in/mikesshapiro/
> [YouTube] @mikesshapiro360
> [TikTok] @mikesshapiro

to learn more about these and much more.

Get going.

Shmalzhaven.

Your brand has to be linear and reflect who you are.

Come on in, the water's fine.

First, get yourself together.

Be self-aware.

There are patterns associated with success, but only you can find your own way.

It takes decades to create great wealth and only ten minutes to unwind it.

Success is you, not what others do.

Never take your wallet to the pit.

Everyone must be accountable to someone.

Nail your pitch.

Don't hold grudges.

Live your clients' lives.

Build your team.

Be "the glue."

Define your vision and values.

Don't coast. You wouldn't drive your car in neutral, so don't drive your life in neutral.

Develop your road map.

Create fortunate timing.

Enjoy yourself.

ABOUT THE AUTHOR

Mike S. Shapiro is CEO of EQTY | Forbes Global Properties and a venture capitalist, entrepreneur, investor, corporate coach, and cofounder and managing director of Plunk, a Seattle-based fintech/proptech analytics company.

Mike's career began as a market maker and trader with the Chicago Board Options Exchange. There, he learned the value of paying attention to what's in front of us—"reading the tape"—and how to use information to predict behaviors, leverage opportunities, and achieve game-changing results. It's the same method he used when he purchased a nearly bankrupt real estate firm in coastal Orange County, California, during the 2008 Great Recession—and that he sold ten years later to create the world's second-largest Sotheby's International Realty brokerage with 1,200 agents and annual sales exceeding $7 billion. He's also used this system to coach 1,000+ real estate agents to greater success, including many who now rank among the "Top 20" agents in the US.

He's a frequent speaker at conferences and universities, a featured guest on television news shows and podcasts, and a sought-after expert on real estate and investing for local and national business publications. His perspectives on real estate and equity markets are available via his weekly blog at mikesshapiro.com.